KERATIN

Edited by **Miroslav Blumenberg**

Keratin

http://dx.doi.org/10.5772/intechopen.73230
Edited by Miroslav Blumenberg

Part of IntechOpen Book Series: Biochemistry, Volume 2
Book Series Editor: Miroslav Blumenberg

Contributors

Lingjuan Zhang, Debananda Ningthoujam, K Tamreihao, Saikat Mukherjee, Rakhi Khunjamayum, Laishram Jaya Devi, Roshan Singh Asem, Xueliang Xiao, Jin-Lian Hu, Ralf Paus, Yuval Ramot, Henry Adeola, Nonhlanhla Khumalo, Jennifer Van Wyk, Afolake Arowolo, Paloma Perez, Lisa Sevilla, Mayumi Komine, Anima Sharma, Tarun Kumar Kumawat, Vishnu Sharma, Subhash Chandra, Miroslav Blumenberg, Sidra Younis

Notice

First published in London, United Kingdom, 2018 by IntechOpen

IntechOpen is the global imprint of INTECHOPEN LIMITED, registered in England and Wales, registration number: 11086078, The Shard, 25th floor, 32 London Bridge Street

London, SE19SG – United Kingdom

Printed in Croatia

British Library Cataloguing-in-Publication Data

A catalogue record for this book is available from the British Library

Additional hard copies can be obtained from orders@intechopen.com

Keratin, Edited by Miroslav Blumenberg

p. cm.

Print ISBN 978-1-78984-804-5

Online ISBN 978-1-78984-805-2

ISSN 2632-0983

IntechOpen Book Series
Biochemistry
Volume 2

Miroslav Blumenberg, PhD., was born in Subotica and received his BSc in Belgrade, Yugoslavia. He completed his PhD at MIT in organic chemistry; this he followed up with two postdoctoral study periods at Stanford University. Since 1983, he has been a faculty member of the RO Perelman Department of Dermatology, NYU School of Medicine, where he is a co-director of a training grant in cutaneous biology. Dr Blumenberg's research is focused on the epidermis, expression of keratin genes, transcription profiling, keratinocyte differentiation, inflammatory diseases and cancers, and most recently, the effects of the microbiome on skin. He has published over 100 peer-reviewed research articles and graduated numerous PhD and post-doctoral students. Dr Blumenberg lives in New York, USA, with his wife and two children.

Book Series Editor and Editor of Volume 2:
Dr. Miroslav Blumenberg
NYU School of Medicine, USA

Scope of the Series

Scope of the Series: Biochemistry, the study of chemical transformations occurring within living organisms, impacts all of life sciences, from molecular crystallography and genetics, to ecology, medicine and population biology. Biochemistry studies macromolecules - proteins, nucleic acids, carbohydrates and lipids –their building blocks, structures, functions and interactions. Much of biochemistry is devoted to enzymes, proteins that catalyze chemical reactions, enzyme structures, mechanisms of action and their roles within cells. Biochemistry also studies small signaling molecules, coenzymes, inhibitors, vitamins and hormones, which play roles in the life process. Biochemical experimentation, besides coopting the methods of classical chemistry, e.g., chromatography, adopted new techniques, e.g., X-ray diffraction, electron microscopy, NMR, radioisotopes, and developed sophisticated microbial genetic tools, e.g., auxotroph mutants and their revertants, fermentation etc. More recently, biochemistry embraced the 'big data' omics systems.Initial biochemical studies have been exclusively analytic: dissecting, purifying and examining individual components of a biological system; in exemplary words of Efraim Racker, (1913 –1991) "Don't waste clean thinking on dirty enzymes." Today however, biochemistry is becoming more agglomerative and comprehensive, setting out to integrate and describe fully a particular biological system. The 'big data' metabolomics can define the complement of small molecules, e.g., in a soil or biofilm sample; proteomics can distinguish all the proteins comprising e.g., serum; metagenomics can identify all the genes in a complex environment e.g., bovine rumen. This Biochemistry Series will address both the current research on biomolecules, and the emerging trends with great promise.

Contents

Preface

Keratin is the proteinaceous body covering layer produced by mammals, birds, fish, reptiles, and amphibians. Importantly, *keratin* is also the intracellular structural protein that protects living epithelial cells from mechanical damage or stress. The fundamental keratin functions are revealed in congenital human skin diseases caused by mutations in keratin genes, e.g., Epidermolysis bullosa simplex or Epidermolytic hyperkeratosis. Most keratin gene mutations have a dominant-negative effect disrupting the filamentous structure formation even from the natural allele, and leaving the cell with a deficient cytoskeleton.

In humans, and mammals in general, keratins are encoded by two large families, the Type I and Type II keratin genes. Obligate heterodimers, keratins are differentially expressed in different epithelia. For example, KRT1 and KRT10 are found in the outermost layers of the skin, KRT3 and KRT12 in the cornea, KRT5 and KRT14 in the basal layer of stratified epithelia, KRT8 and KRT18 in the simple, monolayer epithelia, while KRT6, KRT16, and KRT17 are found in certain proliferative pathological conditions, e.g., psoriasis, during wound healing etc.

The expression of keratins is intricately controlled not only by disease situations, but also by hormones, vitamins, immunomodulatory cytokines, and growth factors. Several chapters in this volume describe the regulation of keratin expression.

Hair and wool, horns, claws, hooves, feathers, and scales are made of keratin. Keratin is insoluble in water and is resistant to proteolysis. However, industrial scale meat production results in vast quantities of keratin byproduct. Processing this byproduct is, on the one hand a major challenge, and on the other hand, a potential for useful recycling and exploitation. Specifically, proteolytic resistance is one major challenge, which necessitates pre-processing of the voluminous keratin bulk using heat and harsh chemicals. On the other hand, keratin contains important amino acids in large quantities; these are useful as feed or as soil enrichment. Several chapters in this volume describe the aspects of keratin exploitation and commercialization.

Miroslav Blumenberg, PhD
Associate Professor of Dermatology
NYU School of Medicine, USA

Section 1

Introduction

Introductory Chapter: Keratins - What to Do with Too Much? What to Do with Too Little?

Miroslav Blumenberg and Sidra Younis

Additional information is available at the end of the chapter

http://dx.doi.org/10.5772/intechopen.79998

1. Introduction

Keratin, from the Greek word for horn, **κέρατο**, denotes the proteinaceous covering layers and structures produced by chordates, including mammals, birds, fish, reptiles, and amphibians. The dead outermost layer of the epidermis, hair and wool, horns, claws, hooves, feathers, and scales is composed of keratin. Keratin is completely insoluble in water and is resistant to proteases that degrade other proteins—1000-year-old Egyptian and other ancient mummies often have full head of hair, virtually undamaged keratin. Keratin proteins can be either alpha-helical in structure, in the skin, hair, and wool of mammals, or parallel sheets of beta-pleated polypeptide chains found in the feathers of birds and scales of reptiles. Rich in amino acid cysteine, keratins become covalently crosslinked via disulfide bonds, which confers a great chemical and biochemical stability to keratin. Thus, keratin serves as important resilient structural and protective functions for the organism.

Importantly, *keratin* is also the resilient structural intracellular protein that protects living epithelial cells from mechanical damage or stress. In cytoplasm, keratin constitutes a filamentous cytoskeletal protein network, extending from the nucleus to the cell periphery, the intermediate filaments, thicker than the actin filaments but thinner than microtubules [1]. Two large families of keratin genes encode multiple proteins with both common and cell-type-specific functions [2].

The indispensable fundamental intracellular keratin functions are revealed in congenital human skin diseases caused by mutations in keratin genes, for example, Epidermolysis bullosa simplex and Epidermolytic hyperkeratosis or in Meesmann's Corneal Dystrophy, the disease caused by a mutation in the gene specifically encoding a corneal keratin [3]. Most keratin gene mutations have a dominant-negative effect, disrupting the filamentous structure formation even from the natural allele and leaving the cell with a deficient cytoskeleton.

2. What to do with too little?

Several chapters in this volume address the diseases associated with keratin deficiencies (see manuscripts by Komine et al., Zhang et al.). Corrective gene therapy approaches attempt to specifically target the mutant keratin gene allele, thus allowing the normal keratin protein to decrease cell fragility [4]. Short inhibitory RNA (siRNA) technology was effectively used to downregulate mutant K6a and K14 allele expressions in cultured PC and EBS cells, respectively [5–7]. This mutation-specific siRNA therapy has been used in a human clinical trial, resulting in effective siRNA treatment of a skin disorder [8]. The functional redundancy of keratins in tissues affected by keratin mutation allows for a possibility to use gene-specific silencing, rather than allele-specific siRNA. Spliceosome-mediated RNA trans-splicing uses the endogenous spliceosome machinery to excise mutant exons and was used to replace the first seven exons of the KRT14 gene in an EBS cell line [9].

Induced pluripotent stem cells and even patient-specific-induced pluripotent stem cells have been generated for use in treatment of inherited keratinopathies [10–12]. Such cell-based therapies have been proposed in conjunction with CRISPR/Cas9- and TALEN-based gene-editing techniques for targeting mutations in the keratin genes [13–15].

A naturally occurring phenomenon, whereby a subpopulation of mutant cells spontaneously reverts to the wild-type phenotype, "revertant mosaicism," has been observed in several patients with EB [16–20]. Revertant mosaicism keratinopathies have two major advantages: (1) the revertant skin is visible and easily accessible and (2) the revertant keratinocytes often have a growth advantage over their mutant progenitors, and so may outgrow and correct the patient's ichthyotic phenotype [21]. Harvesting, expanding, and autologous re-grafting the revertant tissue therefore may be feasible in a clinical setting.

3. What to do with too much?

The increased importance of ecological dangers and ways to alleviate them focused attention on the very large volume of keratin industrial waste. Several chapters in this volume address the incipient remediation efforts (see manuscripts by Ningthoujam et al., Sharma et al., and Nugroho et al.). The mechanical and chemical methodology, cumbersome and inadequate, seems to be giving in to the new biological technology. We can expect deeper understanding of the microbiome and its efficient biodegradation capabilities to play ever more important role. Especially promising are the studies of complex microbiomes, and we can expect in the not-so-distant future that combinations, communities of specific microbes, will be able to convert the obnoxious keratin waste into delightful new materials [22, 23].

Author details

Miroslav Blumenberg[1]* and Sidra Younis[2]

*Address all correspondence to: miroslav.blumenberg@nyumc.org

1 The R. O. Perelman Department of Dermatology, Biochemistry and Molecular Pharmacology, NYU Langone Medical Center, New York, USA

2 Department of Molecular Biology/Biochemistry, National University of Medical Sciences (NUMS), Rawalpindi, Pakistan

References

[1] Loschke F, Seltmann K, Bouameur JE, Magin TM. Regulation of keratin network organization. Current Opinion in Cell Biology. 2015;**32**:56-64. DOI: 10.1016/j.ceb.2014.12.006. PubMed PMID: 25594948

[2] Jacob JT, Coulombe PA, Kwan R, Omary MB. Types I and II keratin intermediate filaments. Cold Spring Harbor Perspectives in Biology. 2018;**10**(4):pii: a018275. DOI: 10.1101/cshperspect.a018275. PubMed PMID: 29610398

[3] Chamcheu JC, Siddiqui IA, Syed DN, Adhami VM, Liovic M, Mukhtar H. Keratin gene mutations in disorders of human skin and its appendages. Archives of Biochemistry and Biophysics. 2011;**508**(2):123-137. DOI: 10.1016/j.abb.2010.12.019. PubMed PMID: 21176769

[4] Cao T, Longley MA, Wang XJ, Roop DR. An inducible mouse model for epidermolysis bullosa simplex: Implications for gene therapy. Journal of Cell Biology. 2001;**152**(3): 651-656. PubMed PMID: 11157990

[5] Dykxhoorn DM, Lieberman J. Knocking down disease with siRNAs. Cell. 2006;**126**(2): 231-235. PubMed PMID: 16873051

[6] Leachman SA, Hickerson RP, Hull PR, Smith FJ, Milstone LM, Lane EB, Bale SJ, Roop DR, McLean WH, Kaspar RL. Therapeutic siRNAs for dominant genetic skin disorders including pachyonychia congenita. Journal of Dermatological Science. 2008;**51**(3):151-157. DOI: 10.1016/j.jdermsci.2008.04.003. PubMed PMID: 18495438

[7] Werner NS, Windoffer R, Strnad P, Grund C, Leube RE, Magin TM. Epidermolysis bullosa simplex-type mutations alter the dynamics of the keratin cytoskeleton and reveal a contribution of actin to the transport of keratin subunits. Molecular Biology of the Cell. 2004;**15**(3):990-1002. PubMed PMID: 14668478

[8] Leachman SA, Hickerson RP, Schwartz ME, Bullough EE, Hutcherson SL, Boucher KM, et al. First-in-human mutation-targeted siRNA phase Ib trial of an inherited skin disorder. Molecular Therapy. 2010;**18**(2):442-446. DOI: 10.1038/mt.2009.273. PubMed PMID: 19935778

[9] Wally V, Brunner M, Lettner T, Wagner M, Koller U, Trost A, Murauer EM, et al. K14 mRNA reprogramming for dominant epidermolysis bullosa simplex. Human Molecular Genetics. 2010;**19**(23):4715-4725. DOI: 10.1093/hmg/ddq405. PubMed PMID: 20861136

[10] Uitto J, Christiano AM, McLean WH, McGrath JA. Novel molecular therapies for heritable skin disorders. The Journal of Investigative Dermatology. 2012;**132**(3 Pt 2):820-828. DOI: 10.1038/jid.2011.389. PubMed PMID: 22158553

[11] Itoh M, Kiuru M, Cairo MS, Christiano AM. Generation of keratinocytes from normal and recessive dystrophic epidermolysis bullosa-induced pluripotent stem cells. Proceedings of the National Academy of Sciences of the United States of America. 2011;**108**(21): 8797-8802. DOI: 10.1073/pnas.1100332108. PubMed PMID: 21555586

[12] Umegaki-Arao N, Pasmooij AM, Itoh M, Cerise JE, Guo Z, Levy B, et al. Induced pluripotent stem cells from human revertant keratinocytes for the treatment of epidermolysis bullosa. Science Translational Medicine. 2014;**6**(264):264ra164. DOI: 10.1126/scitranslmed.3009342. PMID: 25429057

[13] Kocher T, Peking P, Klausegger A, Murauer EM, Hofbauer JP, Wally V, et al. Cut and paste: Efficient homology-directed repair of a dominant negative KRT14 mutation via CRISPR/Cas9 nickases. Molecular Therapy. 2017;**25**(11):2585-2598. DOI: 10.1016/j.ymthe.2017.08.015. PMID: 28888469

[14] Courtney DG, Moore JE, Atkinson SD, Maurizi E, Allen EH, Pedrioli DM, et al. CRISPR/Cas9 DNA cleavage at SNP-derived PAM enables both in vitro and in vivo KRT12 mutation-specific targeting. Gene Therapy. 2016;**23**(1):108-112. DOI: 10.1038/gt.2015.82. PMID: 26289666

[15] Aushev M, Koller U, Mussolino C, Cathomen T, Reichelt J. Traceless targeting and isolation of gene-edited immortalized keratinocytes from epidermolysis bullosa simplex patients. Molecular Therapy—Methods and Clinical Development. 2017;**6**:112-123. DOI: 10.1016/j.omtm.2017.06.008. PMID: 28765827

[16] Kiritsi D, Nanda A, Kohlhase J, Bernhard C, Bruckner-Tuderman L, Happle R, Has C. Extensive postzygotic mosaicism for a novel keratin 10 mutation in epidermolytic ichthyosis. Acta Dermato-Venereologica. 2014;**94**(3):346-348. DOI: 10.2340/00015555-1695. PubMed PMID: 24096702

[17] Smith FJ, Morley SM, McLean WH. Novel mechanism of revertant mosaicism in Dowling-Meara epidermolysis bullosa simplex. The Journal of Investigative Dermatology. 2004; **122**(1):73-77. PubMed PMID: 14962092

[18] Schuilenga-Hut PH, Scheffer H, Pas HH, Nijenhuis M, Buys CH, Jonkman MF. Partial revertant mosaicism of keratin 14 in a patient with recessive epidermolysis bullosa simplex. The Journal of Investigative Dermatology. 2002;**118**(4):626-630. PubMed PMID: 11918708

[19] Lim YH, Fisher JM, Choate KA. Revertant mosaicism in genodermatoses. Cellular and Molecular Life Sciences. 2017;**74**(12):2229-2238. DOI: 10.1007/s00018-017-2468-2. PMID: 28168442

[20] Lai-Cheong JE, McGrath JA, Uitto J. Revertant mosaicism in skin: Natural gene therapy. Trends in Molecular Medicine. 2011;**17**(3):140-148. DOI: 10.1016/j.molmed.2010.11.003. PMID: 21195026

[21] van den Akker PC, Pasmooij AMG, Joenje H, Hofstra RMW, TeMeerman GJ, Jonkman MF. A late-but-fitter revertant cell explains the high frequency of revertant mosaicism in epidermolysis bullosa. PLoS One. 2018;**13**(2):e0192994. DOI: 10.1371/journal.pone.0192994. PMID: 29470523

[22] Toju H, Peay KG, Yamamichi M, Narisawa K, Hiruma K, Naito K, Fukuda S, Ushio M, Nakaoka S, Onoda Y, Yoshida K, Schlaeppi K, Bai Y, Sugiura R, Ichihashi Y, Minamisawa K, Kiers ET. Core microbiomes for sustainable agroecosystems. Nature Plants. 2018;**4**(5):247-257. DOI: 10.1038/s41477-018-0139-4. PubMed PMID: 29725101

[23] Zengler K, Zaramela LS. The social network of microorganisms—How auxotrophies shape complex communities. Nature Reviews. Microbiology. 2018;**16**(6):383-390. DOI: 10.1038/s41579-018-0004-5. PubMed PMID: 29599459

Section 2

Human Skin Keratins

Chapter 2

Regulation of Expression of Keratins and their Pathogenic Roles in Keratinopathies

Mayumi Komine

Additional information is available at the end of the chapter

http://dx.doi.org/10.5772/intechopen.79140

Abstract

Keratins are the epithelia-specific members of intermediate filament superfamily and consist of 54 members. They serve primarily as cytoskeletons, which sustain cell structures. They also influence on cell proliferation and motility by rapidly changing their morphology and distribution through post-translational modification. The expression of keratins genes is regulated by various cytokines and growth factors, mainly through distinct transcription factors. Mutations in keratin genes cause various cutaneous diseases as well as predisposition to inflammatory disorders of internal organ, such as the intestine and the liver. Keratins directly interact signaling molecules, which affects inflammatory processes, and cancer progression. The mechanism of keratin involvement in many diseases will be elucidated in future, which would help identifying novel target for treatment.

Keywords: inflammation, mechanical stress, expression regulation, keratinopathies, keratin

1. Introduction

Keratins are the epithelial-specific members of intermediate filament superfamily, which constitutes the cytoskeleton of cells consisting epithelial tissues, such as stratified epithelia, simple epithelia, hair and nails. Keratin family constitutes with 54 distinct proteins, 28 type I and 26 type II keratins, which expression is tightly regulated in a pairwise fashion. The expression of keratins is site-, differentiation- and context-dependent. Keratin sustains cell-architecture by serving as cytoskeleton, and also it is involved in regulation of cell metabolism and signaling, thereby influencing cell proliferation, migration and apoptosis [1].

Mutation in keratins causes hereditary keratinizing disorders and bullous diseases, such as ichthyosis, palmoplantar keratoderma and epidermolysis bullosa simplex, called "keratinopathies". Mutated keratins cause disruption in cytoskeleton and induce collapse in cell structure. Pathogenic mutations in keratins, which cause epidermolytic ichthyosis and epidermolytic palmoplantar keratoderma, are responsible for hyperkeratosis and inflammation in skin, enhanced by environmental stimuli such as mechanical stress, infection and oxidative stress. Recent studies revealed the role of keratins other than as structural protein in these disorders [2].

Many studies have been performed to reveal the role of keratins in physiological and pathological state, which are far more abundant to follow in this chapter. In this chapter, the role of keratins in physiological state is reviewed and the mutations in keratin genes causing keratinopaties are focused concisely.

2. Keratins

Keratins are the member of intermediate filaments (IF), which composed of six subtypes (**Table 1**) [3]. IFs, 10 nanometer wide filamentous proteins, first described by Holtzer et al. [4] in muscle cells are cytoskeletal proteins constituting almost 70 genes, among which 54 are keratins. Keratins share the structure with other IFs, composed of three domains; a central α-helical rod domain with non-helical head and tail domain containing many phosphorylation sites (**Figure 1**). Keratins have a property of self-assembly, which form filamentous structure. Type I and Type II keratins make heteropolymers, which further form keratin filaments. Type I keratins are acidic, low molecular weight, consists of K9–K40, while type II keratins are basic or neutral, high molecular weight proteins, consists of K1–K8, K71–K86. Expression regulation of each keratin is dependent on tissue-type, differentiation status, and is context-dependent. Keratins are divided into three groups. One is "simple" keratins, expressed in embryonic, and one-layered epithelia, including hepatocytes, intestinal

Subtype	Proteins	Specificity
Type I	Keratins, acidic	Soft stratified epithelia (skin, esophagus, oral mucosa etc.), Soft simple epithelia (gut, sweat gland, etc.), hard epithelia (hair, nail, oral papillae)
Type II	Keratins, basic	
Type III	Vimentin, desmin, glial fibrillary acidic protein, peripherin, syncoilin	Vimentin: mesenchymal cells, desmin: muscle cells, GFAP: astrocytes, glia, peripherin: C-fiber neuron, syncoilin: muscle cells
Type IV	Neurofilament-L, M, H, internexin, synemin, nestin	Neurofilament, internexin: neurons, synemin: muscle cells, nestin: undifferentiated neural cells, neural stem cells
Type V	Lamins A, B, C	Nuclear membrane
Orphan	Filensin, phakinin	Lens

Table 1. Classification of intermediate filament.

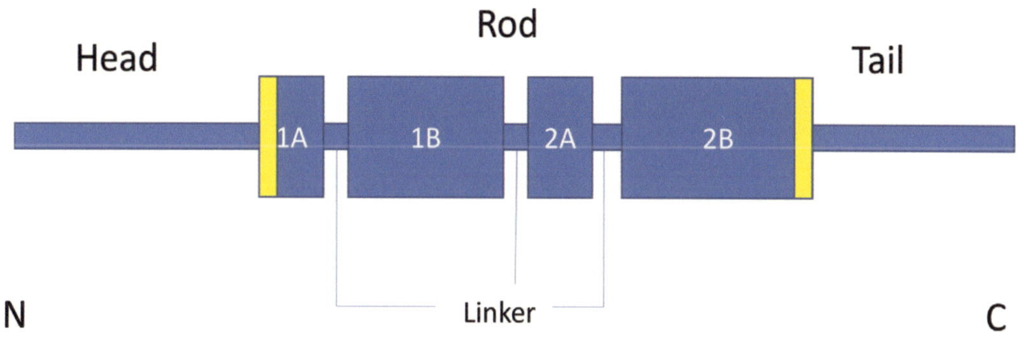

Figure 1. The structure of keratin protein: Central rod domain with head and tail region. Keratin protein consists of rod domain with 4α-helical segments (1A, 1B, 2A, 2B) interconnected with three linker domains, and non-helical head and tail domains. Most of the disease-causing mutations in epidermal keratins occurs in helix initiation and termination motifs at the periphery of 1A domain and 2B domain (indicated with yellow color).

epithelia and sweat glands. "Barrier" keratins are expressed in stratified squamous epithelium, such as skin, oral mucosa and esophagus. "Structural" keratins are hard keratins, constituting hair and nails [3].

Keratin filaments are observed as tonofilaments under electron microscope, which converge at desmosomes and hemidesmosomes. Desmosome is an attachment apparatus between epidermal keratinocytes located at the plasmamembrane of lateral and upper side of basal epidermal keratinocytes, and at all the surrounding plasmamembrane of suprabasal epidermal keratinocytes. Hemidesmosome is an attachment apparatus which conjugates basal keratinocytes to basement membrane, located at the bottom of the basal keratinocytes. These attachment apparatus have a distinct structure. Desmosomes are composed of transmembrane cadherins such as desmogleins, armadillo proteins such as plakoglobins and plakophilins, and plakins such as desmoplakin and plectins which link intracellular desmosomal plaque to keratins (**Figure 2**). Hemidesmosomes composed of integrin α6 and β4, which link hemidesmosomes to laminin, bullous pemphigoid antigen (BPAG)1 and 2, and plectin, which mediate interaction of keratin intermediate filaments to integrins (**Figure 3**). Plectin and desmoplakin anchor keratin filaments to intracellular hemidesmosomal and desmosomal plaque, respectively. Plectin also links keratin filament to nuclear membrane, thus forming cytoskeletal architecture in epidermal keratinocytes [3]. Focal adhesion is another type of adhesion machinery, connecting cells to extracellular matrix (ECM) involving integrins, and anchor actin filaments.

The filament-junction-nucleus network sustains cell structure and rigidity, and anchor cells in three-dimensional architecture of epithelium. The epithelium, however, constantly turnover, regenerates itself when injury, and proliferates at inflammation. Keratin cytoskeleton should be flexible, plastic and dynamic when cells proliferate and migrate. Recent studies revealed how keratin filaments assembles and disassembles in cells. Keratin filament assembly starts at periphery of the cell, close to focal adhesions. In migrating cells, many keratin particles are formed in the lamellipodia where focal adhesions are abundant, which assembles to form keratin filament precursors (KFP). Oligomers of keratin particles are added equally to both ends of KFPs, which become larger in size, and when they approach to the keratin filament

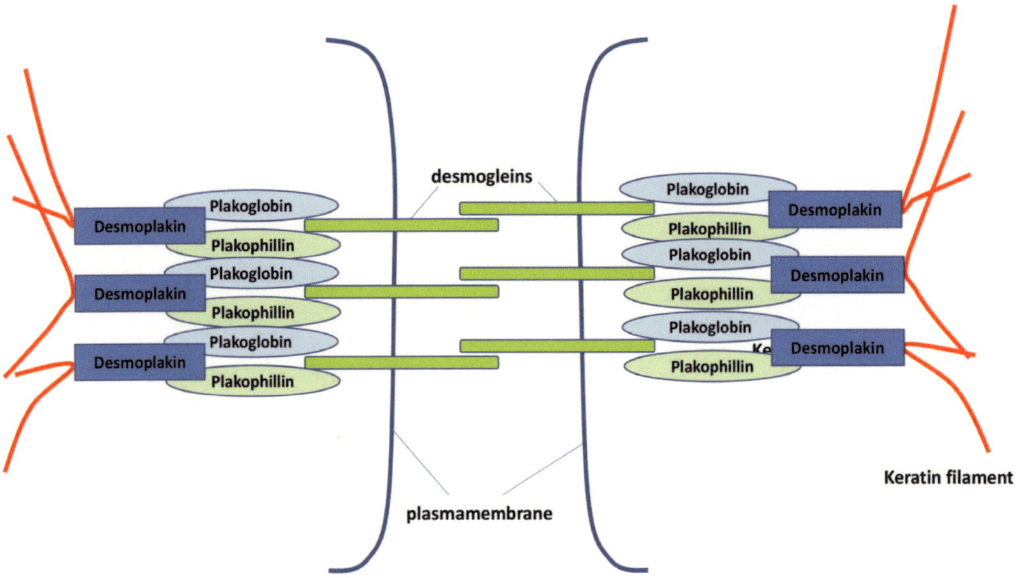

Figure 2. The structure of desmosome.

network close to the nucleus, KFPs integrate in the network adding another branch to the keratin filament network. Keratin filaments further assembles to form bundles close to the nucleus. This assembly of keratin occurs in centripetal flow, while the disassembly occurs close to the nucleus. This "keratin cycle" provides more efficient way over degradation and de novo biosynthesis, and similar recycling system has been observed in other cytoskeletal components, such as actin, and microtubles [5].

Keratins undergo various post-translational modifications, such as phosphorylation, ubiquitination, sumoylation, and acetylation, which regulate the solubility of keratins at specific conditions. Ubiquitination of keratin filaments and subsequent proteasomal degradation have been described as disassembly process. Also, phosphorylation is involved in dissociation of soluble non-filamentous form of keratins. Inhibition of p38 mitogen associated protein kinase (MAPK), or protein kinase C (PKC) ζ results in increased stability, while increased kinase activities result in enhanced keratin filament turnover.

Many phosphorylation sites have been identified on simple epithelial keratins, but fewer in epidermal keratins. Keratin 8 and 18 are phosphorylated when cells are stimulated with sheer stress through protein kinase C(PKC)ζ.

Keratin filament network supports mechanical resilience of the cell, especially for cells of barrier tissues, such as epidermis. Desmosomes and hemidesmosomes, the attachment apparatus of keratinocytes, convey mechanical stress signal to keratins, and to nucleus. When cells become migratory, desmosome-dependent cell–cell adhesion becomes weaker, with reduced co-localization of desmosomes with keratin filaments. Complete knockout of keratin filaments caused scattered distribution of hemidesmosome components, faster attachment to

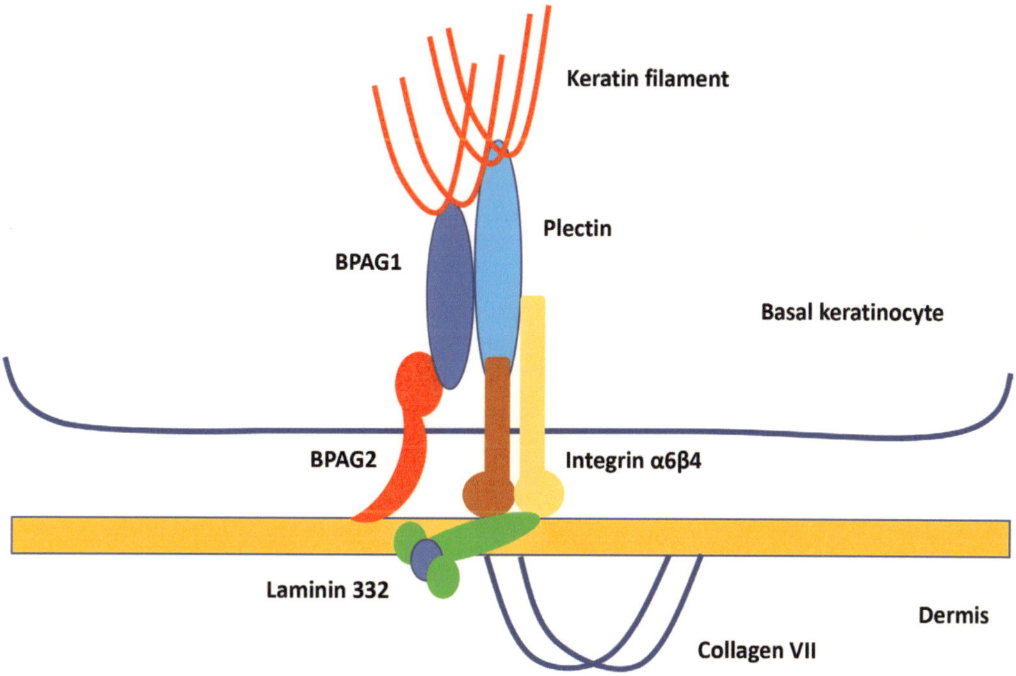

Figure 3. The structure of hemidesmosome. BPAG: Bullous pemphigoid antigen.

extracellular matrix (ECMs), resulting in increased motility. Forced expression of keratin 5 and 14 pairs in the keratin-null keratinocytes, the basal keratinocyte pair of keratins, caused suppressed migration ability. Enhanced migration occurs also in keratinocytes null for epiplakin, plectin, plakoglobin, plakophilin or keratin K6 (K6a/K6b). Loss of keratin K6, plectin or plakoglobin causes activation of Src family kinase, and F-actin reorganization. Src kinase regulates leading edge protrusion though Rac and Cdc42 signaling pathway, which also stimulate to form invadopodia, actin-rich cellular protrusion which works in ECM degradation and cancer invasion and metastasis. Src also directly induces epithelial and mesenchymal transition. Src directly interacts with keratin intermediate filaments in a K6-dependent manner [6]. K6(K6a/K6b) null keratinocytes show enhanced migration, however, K6 is induced in wound healing process where keratinocyte migration is needed. Thus, expression of keratins appears inhibitory to cell migration, which is opposite to expression of vimentin, the type III IF, resulting in enhanced migration and invasion of cancer cells.

3. Regulation of expression of inflammatory keratins

Epidermis covers the outermost surface of human skin. It should withstand the environmental stimuli, such as infection, allergens, and mechanical and chemical insults. Epidermis changes its cytoskeletal keratin expression in inflammatory conditions. Normal healthy interfollicular epidermis expresses keratin K1 and K10 in suprabasal layers, and keratin K5 and K14 in basal layer, while in inflammatory skin conditions, such as psoriasis and atopic dermatitis, expression of keratin K1 and K10 are suppressed and expression of inflammatory keratins, such as keratin K6, K16 and K17 are induced.

Dr. Blumenberg and our group have investigated on the mechanism of induction of these inflammatory keratins. Epidermal growth factor (EGF) induced expression of inflammation and proliferation-related keratins K6 and its counterpart K16 at transcriptional level [7–9]. Interferon (IFN) γ induced keratin K17 promoter activity through transcription factor STAT1 [10, 11]. Tumor necrosis factor (TNF) α induced K6 promoter activity through NFκB and C/EBPβ [12]. Interleukin (IL)-1 induced K6 through C/EBPβ, which binding site clearly distinct from EGF response element [13]. IFNγ also induced keratin K6 through STAT1 signaling pathway [14]. These results indicate that inflammatory cytokines and growth factors induce inflammatory and proliferation-related keratins, at the transcriptional level, which may result in keratinocyte activation in inflammatory skin diseases, such as atopic dermatitis and psoriasis.

Mechanical stimuli are one of important external stimuli, which epidermal keratinocytes respond in daily life. Scratching causes mechanical stretch as well as barrier disruption, and use of tissue expander results in mechanical stretch of epidermal keratinocytes. Pregnant women experiences expansion of abdominal skin caused by growing fetus, especially in third trimester. These mechanical stimuli should influence epidermal keratinocytes through cell–cell junction and cell-ECM junction, such as desmosomes and hemidesmosomes. We have utilized stretchable silicon chamber and examined the effect of mechanical stretch on epidermal keratinocytes. Mechanical stretch induced phosphorylation of EGFR, ERK1/2, and inflammation-related keratin K6, and suppressed differentiation-related keratin K10 [15, 16].

4. Mutation in keratin gene causes various skin diseases and predisposition to internal diseases

Keratin gene mutations causes various diseases, called "keratinopathies" or "keratin disorders". Mutations in epidermal keratins causes skin diseases, such as congenital ichthyosis, congenital bullous disease, and pachyonychia congenita. Mutations in corneal keratins, K3 and K12, cause corneal dystrophy, and mutations in oral keratins, K4/13, cause white sponge nevus in oral mucosa. These causative mutations occur mostly in the conserved region of keratin genes, that is, the beginning/end portion of rod domain, which often affects normal filament assembly, and causes aggregation of keratin protein in the cytoplasm. Mutant keratin affects normal keratin (dominant negative effect), which often results in dominantly inherited congenital diseases. On the other hand, mutation in simple epithelial keratins, such as K8/18 and K19, is found in less conserved regions, which constitutes the risk factors for liver disease and inflammatory bowel diseases.

Mutation in keratin K1 or K10, the differentiation-related epidermal keratins, expressed in the suprabasal layers of epidermis, causes epidermolytic ichthyosis, previously called bullous congenital ichthyosiform erythroderma. The affected child shows erythroderma with bulla formation and later develops ichthyotic skin. The characteristic histological feature is epidermolytic hyperkeratosis, in which hyperkeratosis and coarse keratohyalin granules in degenerated, vacuolar cytoplasm of the granular layer keratinocytes are prominent. The same mutation, when occurred during embryonic development, causes epidermolytic epidermal nevus, aligned in the lines if blaschko, showing similar histological changes in affected skin. Keratin 9 is expressed specifically in the skin of palms and soles. When similar mutations occur in keratin 9, similar histological change, epidermolytic hyperkeratosis, is seen in the epidermis of palms and soles, resulting in Vörner type palmoplantar keratoderma. In these conditions, similar keratin aggregation is observed in the cytoplasm of affected keratinocytes [2, 17].

Keratin 5 and 14 are expressed in pair in the basal layer keratinocytes. Mutation in keratin 5 or 14 causes epidermolysis bullosa simplex, one of congenital bullous disorder. Disruption of keratin filament network in basal keratinocytes with keratin mutation results in collapse of basal keratinocytes, leading to bulla formation at the bottom of the epidermis. Epidermolysis bullosa simplex is one of congenital bullous disease, caused by mutation in basal cell keratin, K5 and K14. Mutated keratin causes fragility in basal keratinocytes where keratin K5 and K14 are expressed, results in intraepidermal bulla formation. Existence of keratin mutations or the reduced expression of keratins cause reduction of desmosome expression and cytoskeletal linker protein expression, resulting in increased motility of keratinocytes. The reduced expression of desmosomes, the junction proteins, may be another mechanism of tissue fragility in EBS patients. Mutations in K5 or K14 in EBS also cause alteration in cellular response to external stress. Keratinocytes with K5 or K14 mutation show increased activation of stress-activated protein kinase (SAPK) signaling against external stresses, as well as constitutive activation of extracellular-signal-regulated kinases (ERKs) [1, 2].

Pigmentary disorder, called Dowling-Degos disease (DDD), has been disclosed to be due to mutation in keratins K5 or K14, and Galli-Galli disease (GGD) due to mutation in keratin K5. DDD patients classically show small pigmented macules and reticulated pigmentation in the large folds and flexure surface. Some patients show pigmentation on the face, and also comedo-like papules and pitted scarring are seen. GGD patients show similar clinical features, with distinct histological changes, including acantholysis. Patients with EBS with mottled pigmentation show similar pigmentary changes with vesicle formation. Mutations in keratin K5 have been reported to cause EBS since the 1990s; there is a genotype–phenotype correlation between keratin mutation and the type of EBS. The most severe form, Dowling-Meara type, is caused by mutation in the highly conserved region on the either side of the helix boundary area of rod domain, while the milder type of EBS, Koebner type, and Weber-Cockayne type are caused by mutations occurred throughout the rod domain. There is a subtype of EBS with mottled pigmentation caused by mostly specific mutation p.Pro25Leu, in the head domain of keratin K5. Mutations in K5 of DDD and DDG are also in the head domain of K5, which are nonsense mutation or frameshift mutation, resulting in premature stop codons, leading to haploinsufficiency of K5 rather than dominant negative effect. These cases demonstrate that keratin K5 is important in melanosome transportation. Melanosomes, one of cell organelles containing melanin, are produced in melanocytes and transferred to keratinocytes where melanosomes distribute in the cytoplasm. Mutation in K5 results in melanosome transfer and the distribution of melanosomes in keratinocytes, leading to altered pigmentation in skin. Thus, keratins may also contribute in organelle transfer in keratinocytes [1].

Keratin K6, K16, and K17 are expressed in follicular epithelium, oral mucosa, palms and soles and nails. They are also induced in inflammatory skin diseases in interfollicular epidermis. Mutation in keratin K6, K16 or K17 causes pachyonychia congenita, showing thick and deforming nails and hyperkeratotic palms and soles, with or without steatocystoma multiplex. Pachyonychia congenita (PC)-1, is a form of PC presenting with nail defects, palmoplantar hyperkeratosis, follicular hyperkeratosis, and oral leukokeratosis. PC-2 lacks oral involvement, but has multiple folliculosebaceous cysts and natal teeth. Kratin 6 consists of three isoforms, K6A, K6B and K6C. K6A is the most abundant isoform, which makes pair with K16, and the mutation in K6A, as well as K16 leads to PC-1. K6B is a counterpart of K17, and mutation in K6B or K17 leads to PC-2 [2, 17]. These keratins, K6, K16 and K17 are induced by several cytokines and growth factors, such as EGF, TNF and IFNγ, or by mechanical stress or UV. The attempt to suppress the expression of mutated keratins has been done by several researchers to treat PC patients. Small interfering RNA for mutant K6a has been tried to treat PC patients, demonstrating feasible therapeutic strategy for keratin disorders [18]. RNA interference, however, harbors potential risk for off-target effects, which should be effectively avoided. Recently, K16 has been disclosed to be involved in the induction of danger-associated molecular patterns (DAMPs)/alarmins and skin barrier genes [19], and the regulator of nuclear factor erythroid-derived 2 related factor 2 (NRF2). In PC lesional epidermis, NRF2 protein expression is elevated, but the activation of NRF2 is suppressed, similar to Krt16−/− mice which present PC-like skin lesions. Reduction in active NRF2 results in reduced glutathione (GSH) levels, indicating increased oxidative stress. Inhibitor of GSH synthesis has been shown to induce PPK in mice, which imply that reduced NRF2 activity in Krt16−/− mice caused PPK through reduction in GSH levels and increased oxidative stress. Topical application of sulforaphane, an activator of NRF2, rescued PPK in Krt16−/− mice, as well as increasing the levels of NRF2, and pNRF2, indicating the possibility of treating PPK with small molecule targeted drugs pharmacologically activating NRF2 [20].

Mutations in K8 or K18, simple epithelial keratins, have been shown to be the risk factor for some patients with inflammatory bowel diseases or liver disease. Mutation in cytoskeletal keratins causes reduced resilience in epithelia of digestive tract which is always under mechanical stress and peristaltic movement, resulting in cellular damage. Mutations in keratin in liver cause reduced tolerance to toxins, such as alcohol and drugs, thus predispose patients to liver damage at situations of cell stress. Keratin K8 undergoes hyperphosphorylation, acting as phosphate sponge to absorb various phosphorylated proteins, such as SAPK and ERKs, and reduce inflammation and apoptosis [2, 17].

5. Keratins and inflammatory diseases and cancer

Keratin filaments regulate inflammatory processes. Wild-type keratin K8 has been reported to be a negative regulator of inflammation by suppressing TLR signaling through inhibiting

NFκB activation [21]. Keratin 8 also protects colonic epithelium from inflammation, and cancer progression [22, 23] Wild-type keratins suppress TSLP production. Normal human epidermal keratinocytes do not produce TSLP in culture, while keratin null keratinocytes produce copious amount of TSLP [24]. They showed that defects in keratins caused activation of MEK1/2 and ERK1/2, resulting in TSLP production independent on barrier disruption. Keratin 17 expression promotes inflammation towards Th1- and Th17-type immune reactions, the characteristic inflammation in psoriasis, and the absence of K17 attenuates inflammation and tumorigenesis [25]. C-terminus of K6 has anti-microbial properties [26]. Mice with K1 expression in pancreas βcells develop diabetes with decreased insulin secreting vesicles. K17 binds to 14-3-3σ and TRADD, and loss of K17 results in altered inflammatory cytokine production, impaired wound healing and impaired hair follicle cycling [2]. K14 also directly binds to TRADD, which influences the signaling pathway through TNFR [27]. K16 knockout mice show increased activity of danger signals, described as above. Another report showed that the amount of keratin protein itself is important to protect cells against mechanical stress [28].

Keratins are also involved in cancer proliferation and invasion. Keratin 19 fragment (CYFRA) has been used in clinics for tumor marker, especially to detect non-small cell lung cancer. Not only as a tumor marker, K19 promotes tumor cell invasion in hepatocellular carcinoma, probably by formation of invadopodia [29]. Cancer cell with K19 expression also shows increased resistance to chemotherapy [1].

6. Conclusion

Keratins are cytoskeletal proteins, however, not only that, keratins have various roles in physiology and pathophysiology of human organs, and involved in proliferation, motility and invasion of the cells, and inflammation of tissues. Their complexed behavior would be further elucidated in future, and many more novel findings would help exploring the target of future therapy of inherited cutaneous diseases, cancer and inflammatory disorders.

Conflict of interest

None.

Author details

Mayumi Komine

Address all correspondence to: mkomine12@jichi.ac.jp

Department of Dermatology, Jichi Medical University, Shimotsuke, Tochigi, Japan

References

[1] Pan X, Hobbs RP, Coulombe PA. The expanding significance of keratin intermediate filaments in normal and diseased epithelia. Current Opinion in Cell Biology. Feb 2013;**25**(1):47-56

[2] Haines RL, Lane EB. Keratins and disease at a glance. Journal of Cell Science. Sep 1, 2012;**125**(Pt 17):3923-3928

[3] Chung BM, Rotty JD, Coulombe PA. Networking galore: Intermediate filaments and cell migration. Current Opinion in Cell Biology. 2013;**25**:600-612

[4] Ishikawa H, Bischoff R, Holtzer H. Mitosis and intermediate-sized filaments in developing skeletal muscle. Journal of Cell Biology. 1968;**38**:538-555

[5] Windoffer R, Beil M, Magin TM, Leube RE. Cytoskeleton in motion: The dynamics of keratin intermediate filaments in epithelia. Journal of Cell Biology. 2011;**194**:669-178

[6] Loschke F, Seltmann K, Bouameur JE, Magin TM. Regulation of keratin network organization. Current Opinion in Cell Biology. 2015;**32**:56-65

[7] Jiang CK, Magnaldo T, Ohtsuki M, Freedberg IM, Bernerd F, Blumenberg M. Epidermal growth factor and transforming growth factor alpha specifically induce the activation- and hyperproliferation-associated keratins 6 and 16. Proceedings of the National Academy of Sciences of the United States of America. Jul 15, 1993;**90**(14):6786-6790

[8] Bernerd F, Magnaldo T, Freedberg IM, Blumenberg M. Expression of the carcinoma-associated keratin K6 and the role of AP-1 proto-oncoproteins. Gene Expression. 1993;**3**(2):187-199

[9] Magnaldo T, Bernerd F, Freedberg IM, Ohtsuki M, Blumenberg M. Transcriptional regulators of expression of K#16, the disease-associated keratin. DNA and Cell Biology. Dec 1993;**12**(10):911-923

[10] Jiang CK, Flanagan S, Ohtsuki M, Shuai K, Freedberg IM, Blumenberg M. Disease-activated transcription factor: Allergic reactions in human skin cause nuclear translocation of STAT-91 and induce synthesis of keratin K17. Molecular and Cellular Biology. Jul 1994;**14**(7):4759-4769

[11] Komine M, Freedberg IM, Blumenberg M. Regulation of epidermal expression of keratin K17 in inflammatory skin diseases. The Journal of Investigative Dermatology. Oct 1996;**107**(4):569-575

[12] Komine M, Rao LS, Kaneko T, Tomic-Canic M, Tamaki K, Freedberg IM, Blumenberg M. Inflammatory versus proliferative processes in epidermis. Tumor necrosis factor alpha induces K6b keratin synthesis through a transcriptional complex containing NFkappa B and C/EBPbeta. The Journal of Biological Chemistry. Oct 13, 2000;**275**(41):32077-32088

[13] Komine M, Rao LS, Freedberg IM, Simon M, Milisavljevic V, Blumenberg M. Interleukin-1 induces transcription of keratin K6 in human epidermal keratinocytes. The Journal of Investigative Dermatology. Feb 2001;**116**(2):330-338

[14] Hattori N, Komine M, Yano S, Kaneko T, Hanakawa Y, Hashimoto K, Tamaki K. Interferon-gamma, a strong suppressor of cell proliferation, induces upregulation of keratin K6, one of the inflammatory- and proliferation-associated keratins. The Journal of Investigative Dermatology. Aug 2002;**119**(2):403-410

[15] Yano S, Komine M, Fujimoto M, Okochi H, Tamaki K. Mechanical stretching in vitro regulates signal transduction pathways and cellular proliferation in human epidermal keratinocytes. The Journal of Investigative Dermatology. Mar 2004;**122**(3):783-790

[16] Yano S, Komine M, Fujimoto M, Okochi H, Tamaki K. Activation of Akt by mechanical stretching in human epidermal keratinocytes. Experimental Dermatology. May 2006;**15**(5):356-361

[17] Toivola DM, Boor P, Alam C, Strnad P. Keratins in health and disease. Current Opinion in Cell Biology. Feb 2015;**32**:73-81

[18] Leachman SA, Hickerson RP, Schwartz ME, Bullough EE, Hutcherson SL, Boucher KM, Hansen CD, Eliason MJ, Srivatsa GS, Kornbrust DJ, Smith FJ, McLean WI, Milstone LM, Kaspar RL. First-in-human mutation-targeted siRNA phase Ib trial of an inherited skin disorder. Molecular Therapy. Feb 2010;**18**(2):442-446

[19] Lessard JC, Piña-Paz S, Rotty JD, Hickerson RP, Kaspar RL, Balmain A, Coulombe PA. Keratin 16 regulates innate immunity in response to epidermal barrier breach. Proceedings of the National Academy of Sciences of the United States of America. Nov 26, 2013;**110**(48):19537-19542

[20] Kerns ML, Hakim JM, Lu RG, Guo Y, Berroth A, Kaspar RL, Coulombe PA. Oxidative stress and dysfunctional NRF2 underlie pachyonychia congenita phenotypes. The Journal of Clinical Investigation. Jun 1, 2016;**126**(6):2356-2366

[21] Dong XM, Liu ED, Meng YX, Liu C, Bi YL, Wu HW, Jin YC, Yao JH, Tang LJ, Wang J, Li M, Zhang C, Yu M, Zhan YQ, Chen H, Ge CH, Yang XM, Li CY. Keratin 8 limits TLR-triggered inflammatory responses through inhibiting TRAF6 polyubiquitination. Scientific Reports. Sep 2, 2016;**6**:32710

[22] Habtezion A, Toivola DM, Butcher EC, Omary MB. Keratin-8-deficient mice develop chronic spontaneous Th2 colitis amenable to antibiotic treatment. Journal of Cell Science. May 1, 2005;**118**(Pt 9):1971-1980 [Epub Apr 19, 2005]

[23] Liu C, Liu ED, Meng YX, Dong XM, Bi YL, Wu HW, Jin YC, Zhao K, Li JJ, Yu M, Zhan YQ, Chen H, Ge CH, Yang XM, Li CY. Keratin 8 reduces colonic permeability and maintains gut microbiota homeostasis, protecting against colitis and colitis-associated tumorigenesis. Oncotarget. May 27, 2017;**8**(57):96774-96790

[24] Kumar V, Behr M, Kiritsi D, Scheffschick A, Grahnert A, Homberg M, Schwieger-Briel A, Jakob T, Bruckner-Tuderman L, Magin TM. Keratin-dependent thymic stromal lymphopoietin expression suggests a link between skin blistering and atopic disease. The Journal of Allergy and Clinical Immunology. Nov 2016;**138**(5):1461-1464.e6

[25] Depianto D, Kerns ML, Dlugosz AA, Coulombe PA. Keratin 17 promotes epithelial proliferation and tumor growth by polarizing the immune response in skin. Nature Genetics. 2010;**42**:910-914

[26] Tam C, Mun JJ, Evans DJ, Fleiszig SM. Cytokeratins mediate epithelial innate defense through their antimicrobial properties. The Journal of Clinical Investigation. Oct 2012; **122**(10):3665-3677

[27] Yoneda K, Furukawa T, Zheng YJ, Momoi T, Izawa I, Inagaki M, Manabe M, Inagaki N. An autocrine/paracrine loop linking keratin 14 aggregates to tumor necrosis factor alpha-mediated cytotoxicity in a keratinocyte model of epidermolysis bullosa simplex. The Journal of Biological Chemistry. Feb 20, 2004;**279**(8):7296-7303

[28] Asghar MN, Silvander JS, Helenius TO, Lähdeniemi IA, Alam C, Fortelius LE, Holmsten RO, Toivola DM. The amount of keratins matters for stress protection of the colonic epithelium. PLoS One. May 22, 2015;**10**(5):e0127436. DOI: 10.1371/journal.pone.0127436. eClloction 2015

[29] Govaere O, Komuta M, Berkers J, Spee B, Janssen C, de Luca F, Katoonizadeh A, Wouters J, van Kempen LC, Durnez A, et al. Keratin 19: A key role player in the invasion of human hepatocellular carcinomas. Gut. 2014;**63**:674-685

Chapter 3

Glucocorticoid Receptor Signaling in Skin Barrier Function

Lisa M. Sevilla and Paloma Pérez

Additional information is available at the end of the chapter

http://dx.doi.org/10.5772/intechopen.79458

Abstract

Glucocorticoids (GCs) are steroid hormones that regulate the physiology of all tissues and mediate stress responses. Synthetic GCs are commonly prescribed to treat chronic inflammatory conditions including the prevalent skin diseases—psoriasis and atopic dermatitis. GCs act through the GC receptor (GR, NR3C1), a ligand-activated transcription factor belonging to the nuclear hormone receptor superfamily. In skin, GC therapeutic efficacy is due to the antiproliferative and anti-inflammatory actions of GR; however, in the long term, these benefits are accompanied by adverse profiles including skin atrophy, increased fragility, dehydration, augmented susceptibility to infections, and delayed wound healing. While the therapeutic actions of GC treatments have been extensively studied, only more recently has the physiological role of GR been addressed in skin. *In vivo* and *in vitro* studies in mouse and man have revealed an important function for GR in skin homeostasis. In particular, the characterization of gain- or loss-of-function mouse models has demonstrated relevant roles for GR in skin pathophysiology. The actions of GR are context dependent, and in skin, it regulates different gene subsets and biological processes depending on developmental stage and physiological state. Finally, recent findings emphasize the relevance of local GC biosynthesis and appropriate GR expression in maintaining skin homeostasis.

Keywords: glucocorticoids, glucocorticoid receptor, skin barrier, keratins, transcriptional regulation, keratinocyte proliferation and differentiation, inflammatory skin diseases

1. Introduction

Endogenous glucocorticoids (GCs) are steroid hormones that regulate a vast array of biological processes, including development, cellular proliferation and differentiation, metabolism, immunity, and stress response [1, 2]. In response to physiological cues and stressors, the hypothalamic-pituitary-adrenal (HPA) axis coordinates the systemic production and secretion of GCs from the zona fasciculata of the adrenal glands. Importantly, extra-adrenal, local synthesis of GCs occurs in multiple tissues including the thymus, intestine, brain, and skin, which express functional equivalents of the HPA axis [3, 4]. The tissue availability of these steroid hormones is further regulated by locally expressed 11β-hydroxysteroid dehydrogenase type I and II enzymes (HSD11B1 and 2), which catalyze the interconversion between active (cortisol and corticosterone) and inactive (cortisone and 11-dehydrocorticosterone) forms in humans and rodents, respectively [5].

The GC receptor (GR) is a ubiquitously expressed ligand-dependent transcription factor (TF) that belongs to the nuclear receptor (NR) superfamily and mediates the physiological and pharmacological actions of GCs [2, 6]. The actions of GR on transcription are highly context specific, with strikingly different subsets of genes being regulated across different cell types, developmental stages, and pathophysiological states [7].

For more than half a century, synthetic GCs have been used clinically to manage autoimmune diseases due to their potent anti-inflammatory and immunosuppressive properties [8]. However, their therapeutic use is limited by a host of undesired side effects ranging from osteoporosis, obesity, and muscle wasting to skin atrophy and impaired wound healing [9]. Nevertheless, GCs are still the most effective and widely prescribed therapeutic agent for prevalent inflammatory skin diseases including atopic dermatitis, with a lifetime prevalence of 10–20% in developed countries; and psoriasis, affecting 2% of the European and North American population [10, 11]. Keratins, comprising approximately 30% of epidermal proteins and more than 90% in hair follicles, play key roles in maintaining skin barrier function. In fact, keratin mutations are associated to many genodermatoses and their expression is altered upon inflammation, wounding, or tissue damage. Hormones, and in particular GCs, are major regulators of keratin gene expression in healthy and diseased skin [12].

1.1. The glucocorticoid receptor: structure and function

GR was the first identified member of the NR superfamily and characteristic of this group; its domains comprise an N-terminal transactivation domain, a central DNA-binding domain, and a C-terminal ligand-binding domain (**Figure 1A**) [2, 6]. The N-terminal domain is responsible for interactions with the transcriptional machinery as well as coregulators via the activation function (AF)-1 region. The majority of sites for posttranslational modification, including phosphorylation, ubiquitination, and sumoylation are located in this domain, allowing for modulation of receptor function and contributing to context specificity [2, 6]. The highly conserved DNA-binding domain has two zinc finger motifs responsible for recognizing and binding to GC response elements (GREs) as well as a nuclear localization signal. A flexible hinge region connects the DNA-binding domain with the C-terminal ligand-binding domain that contains a hydrophobic pocket for GC binding, an AF-2 region for ligand-dependent interactions with coregulators and a second nuclear localization signal. Receptor dimerization is mediated by sequences in the DNA- and ligand-binding domains [6, 13].

The gene encoding GR, *NR3C1*, contains nine exons with the open reading frame being encoded by exons 2–9. The *NR3C1* transcript can undergo alternative splicing resulting in the

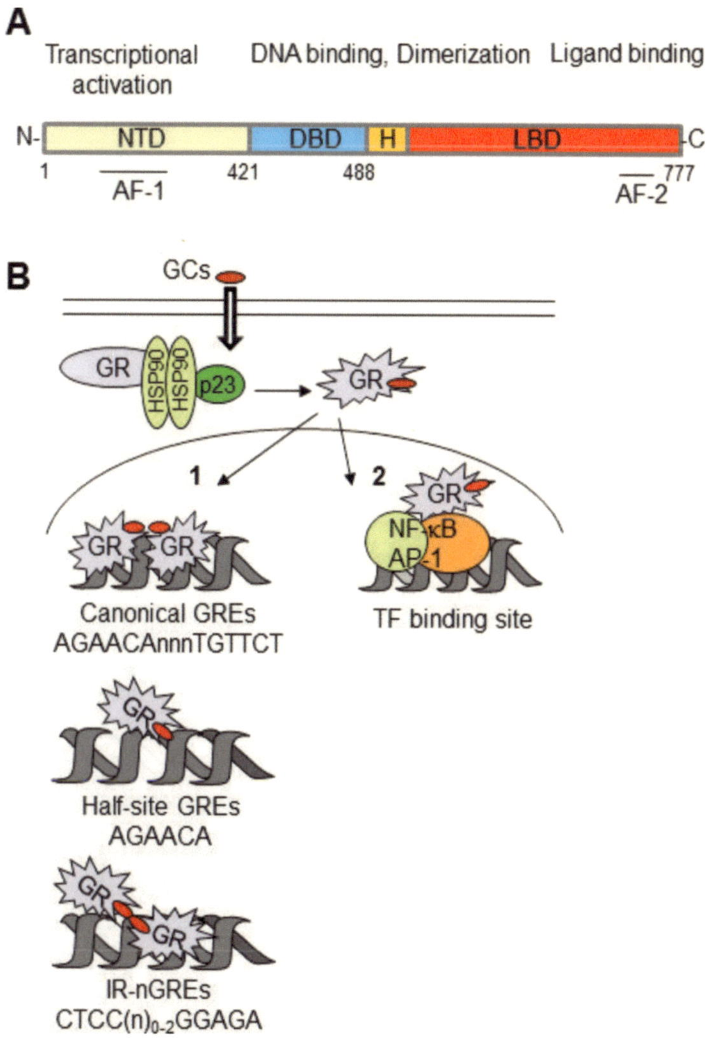

Figure 1. Glucocorticoid (GC) signaling through the GC receptor (GR). (A) Functional domains of the GR. NTD, N-terminal transactivation domain; DBD, DNA-binding domain; H, hinge; LBD, ligand-binding domain; AF, activation function. (B) Scheme of GC-regulated transcription. In the absence of ligand, GR is mostly associated to cytoplasmic multiprotein complexes including chaperones. Upon GC binding, activated GR is released from this complex, dimerizes, translocates to the nucleus, and binds directly to DNA (1) or to other TFs via tethering (2) to regulate gene expression. GR-bound genomic regions include canonical GRE elements (imperfect palindrome GREs), half-site GREs, and inverted repeat (IR)-negative (n)GREs.

generation of at least five isoforms [2], with GRα and GRβ, differing at their C-termini, being the most studied. GRα binds to ligand and carries out classical receptor functions, and will be referred to as simply GR elsewhere in this chapter. On the other hand, GRβ is incapable of binding to GCs and acts as a dominant-negative inhibitor of GRα. GRβ is usually expressed at a lower level than GRα, but alterations in the ratio between these isoforms are associated with GC sensitivity and autoimmune disease [2]. The expression of these splice variants in healthy and diseased skin is only beginning to be explored. For example, increased GRβ was found in patients with severe atopic dermatitis that were unresponsive to GC treatment [14]. However, a more recent study failed to establish a correlation between isoform ratio and sensitivity to GC therapies in inflammatory dermatoses [15]. Another layer of complexity is added by the discovery that seven GR protein isoforms are generated through alternative translational initiation [2]. These isoforms are differentially expressed across tissues, and while all are capable of binding DNA and ligand, the N-terminal truncations do alter subcellular localization and transcriptional activity. Their precise role in skin pathophysiology remains to be determined.

In the absence of ligand, the majority of GR is sequestered in the cytoplasm in a multiprotein complex that includes chaperones (HSP70, HSP90, and p23) and immunophilins. The classical model of GR activation is that upon binding to GCs, GR dissociates from this complex, dimerizes, translocates to the nucleus, and binds to GREs or to other TFs, regulating gene expression (**Figure 1B**). GR-bound genomic regions are widespread throughout the genome, and are not necessarily found in close proximity to target genes [16, 17]. The canonical GRE sequence is 5′-AGA ACA nnn TGT TCT-3′, an imperfect palindrome that contains two half sites and a three base pair spacer. This classical mechanism of transcriptional regulation, which is dependent on DNA-binding and dimerization, was denominated as transactivation. In contrast to other TF-binding sites, GREs show a great deal of variability, with changes in the majority of positions not impeding GR binding [16]. Remarkably, the very sequence of the GRE was demonstrated to affect GR conformation and transcriptional activity, functioning as an allosteric regulator of this TF [18]. Classical GREs are not the only mode for GR chromatin interaction, as GR can also regulate transcription by binding to half site sequences [19], such as those found near the epidermal keratin (K)5, 14, 6, and 17 genes [20]. These keratin response elements allow the simultaneous binding of at least two NRs, which widens the hormone-dependent transcriptional control of keratin expression [21].

Finally, the most recently identified GR-bound regulatory sequence is the inverted repeat (IR)-negative (n) GRE with the consensus sequence 5′-CTCC $(n)_{0-2}$ GGAGA-3′ [22]. These sites are termed negative as they promote the assembly of cis-acting corepressor complexes that recruit histone deacetylases resulting in gene repression. Other mechanisms by which GR modulates transcription are by binding composite elements, or juxtaposed binding motifs, with other TFs; and by modulating the transcriptional activity of other TFs by protein-protein interactions, independent of DNA-binding and receptor dimerization, known as tethering (**Figure 1B**) [23]. It was long assumed that tethering, and in particular GR interference with prototypical proinflammatory TFs such as NF-κB and AP-1, mediated the beneficial anti-inflammatory actions of GR, while transactivation was responsible of the adverse side effects. This was mostly based on studies using a GR single mutant (A458T), which impeded dimerization and impaired transactivation of GRE-containing target genes while allowing tethering via AP-1 and NF-κB. This dogma was recently challenged by demonstrating that GR^{A458T} could indeed dimerize and bind DNA in a subset of GREs in live cells although with reduced efficiency [13].

More than 20 GR chromatin immunoprecipitation sequencing (ChIP-Seq) studies in different cell/tissue types have been published thus far [7], providing functional insights. For instance, GR largely relies upon other factors to create and maintain open chromatin, contributing to its context specificity [7]. Thus far, the only GR ChIP-Seq experiment performed in keratinocytes evaluated a short treatment with dexamethasone (Dex) [24]. Following a restrictive analysis, 104 GR-bound genomic sites were identified. This small number contrasts with the thousands of targets identified in other cell types [7], however, is in line with transcriptomic data following 4 h of GC treatment in primary human keratinocytes [25]. Despite the limited number of GR-bound genomic sites detected, their analysis provided important information about functional interactions between GR and other TFs in keratinocyte gene regulation, as several overrepresented TF motifs were identified, including KLF (43%) and AP-1 (28%) [24]. Further experimentation revealed that GR and KLF4 cooperate to regulate the expression of the anti-inflammatory genes *Gilz/Tsc22d3* (GC-induced leucine zipper) and *Zfp36/Tristetraprolin*.

Ligand binding of GR also results in rapid actions, occurring within seconds to minutes, which occur independently of transcription or translation, commonly referred to as nongenomic actions [9]. For instance, ligand-bound GR interferes with the phosphatidylinositol-3-kinase signaling pathway and the downstream kinase AKT, critical for cell proliferation and survival. This interference has been demonstrated in mouse skin and cultured keratinocytes and contributes to the antitumor effects of GCs in this tissue [26].

1.2. Adrenal and cutaneous GC production

GC signaling represents a complex homeostatic system that mediates fundamental tissue-specific processes during development as well as adaptive responses to stress. The importance of appropriate GC levels for normal tissue function is clearly illustrated in extreme situations of hormone imbalances where chronic excess or deficiency leads to pathological conditions, such as Cushing's or Addison's disease, respectively [1]. In both scenarios, dysfunctional responses to this hormone can result in differential tissue sensitivity and manifest as clinical GC resistance (e.g., primary generalized glucocorticoid resistance) or hypersensitivity [27]. Remarkably, cutaneous abnormalities in Cushing's patients—skin atrophy, increased fragility and easy bruising, elevated infection risk, and impaired wound healing—are very similar to those found in aging and also after long-term/high-dose GC pharmacological treatments [28, 29].

As GC synthesis and release is tightly controlled by the HPA axis, this neuroendocrine system acts as a major regulator of skin integrity and function. Stress or physiological conditions stimulate the production of corticotropin-releasing hormone receptor (CRH) from the hypothalamus, which in turn induces the release of adrenocorticotropic hormone (ACTH) from the anterior pituitary, which promotes the synthesis and release of GCs from the adrenal cortex. Under physiological conditions, GC release shows a diurnal pattern controlled by the circadian clock, with peak levels linked to the beginning of daily activity. GCs can inhibit their own production by a feedback mechanism where GR shuts off the secretion of CRH and ACTH [1, 2]. Around 90% of circulating cortisol is bound with high affinity by corticosteroid-binding globulin, while the remaining 10% of circulating GCs consist of roughly equal proportions of cortisol and cortisone. The free circulating cortisone functions as a reservoir of inactive steroid that can be converted into active GCs in a tissue-specific manner.

The discovery that the skin behaves as a local HPA axis analog that is able to produce steroidogenic enzymes and GCs constituted a major breakthrough for understanding alternative mechanisms by which GCs exert their actions in homeostatic and pathological conditions [3, 30]. Since skin is continuously exposed to external perturbations, tissue-specific synthesis of GCs represents an ideal response mechanism and recent studies suggest that systemic and local HPA axes are interconnected [31]. Recent findings emphasize the relevance of the local GC biosynthetic pathway in maintaining skin homeostasis as local GC deficiency and reduced GR expression in psoriatic lesions contributed to the pathogenesis of the disease. These findings should be considered for designing novel GC-based strategies for treating skin diseases [32, 33].

HSD11B1/HSD11B2 activities maintain appropriate GC levels and constitute a key mechanism to modulate GR function at the prereceptor level both in plasma and in peripheral tissues [5]. HSD11B2 expression and activity is key in the renal and cardiovascular system where GC inactivation is required to avoid the overactivation of the closely related mineralocorticoid receptor, favoring instead binding of the mineralocorticoid aldosterone [34, 35]. In human and mouse skin, HSD11B1 is highly expressed in the epidermis and dermis, with higher levels in differentiating keratinocytes [29, 36]. HSD11B2 has been also detected in the suprabasal epidermis of human and developing mouse skin as well as in sweat glands, an important target for aldosterone-mineralocorticoid receptor regulation [37–40].

2. GR function in skin development

2.1. Development of the epidermis and its appendages

Barrier formation begins with epidermal commitment around E10.5 when surface ectoderm cells begin to express the keratinocyte-specific intermediate filament proteins K5 and 14 [41]. By E14.5, keratinocytes stratify, express K1 and 10, and begin terminal differentiation forming the postmitotic spinous and granular layers and the outermost SC [42]. The SC is composed of fully differentiated dead keratinocytes, or corneocytes (described as bricks) surrounded by specialized extracellular lipids (or mortar), extruded by lamellar bodies at the granular layer-SC interface [43]. Elegant studies subjecting mouse embryos to whole mount dye exclusion assays revealed that the epidermal permeability barrier acquisition is patterned, beginning at initiation sites at ~E16.5 and spreading in moving fronts until completion by ~E17.5 [44]. Hair follicle patterning and morphogenesis begins at E14.5 when placodes, or clusters of basal keratinocytes, form stimulated by inductive signals from the dermis [45, 46]. Sebaceous glands begin to form toward the end of gestation, and pilosebaceous units continue maturation postnatally. Eccrine sweat glands begin to form late in mouse development and are restricted to paw pads. All epidermal appendages contribute to skin function, as sweat glands and hair contribute to thermoregulation and sebaceous glands secrete lipid-rich sebum that waterproofs the skin and has antimicrobial activities. Importantly, defects in epidermal differentiation during development can lead to inflammatory skin disease later in life [45].

The signaling pathways orchestrating epidermal development and keratinocyte terminal differentiation have been extensively studied [41, 42, 46]. The master regulator TF p63 is crucial for early epidermal specification and differentiation, but its expression must decrease in keratinocyte terminal differentiation, a process regulated by functional interactions between more than 50 TFs, including GR [41, 47, 48]. A keratinocyte cell line derived from mice deficient in epidermal GR (see Section 3.2) showed defects in terminal differentiation, with an increased expression of the predominant isoform ΔNp63 [24]. Further experimentation showed that GR inhibits p63 expression, fitting with its proposed role as an inhibitor of the early stages of keratinocyte differentiation [24, 25]. Heterozygous mutations are present in the *TRP63* gene in patients with different ectodermal dysplasias, developmental disorders in which the epidermis and its appendages fail to develop normally. Mouse models with reduced expression of p63 mimic features of the human disease [47]. Strikingly, transgenic mice with ectodermal overexpression of GR also exhibit features of ectodermal dysplasia, strong evidence of functional interactions between these TFs [49, 50].

2.2. GCs and skin barrier formation

During development, GCs are provided maternally as well as by the embryo; in mice, systemic synthesis begins around E14 and peaks around birth [51]. Embryonic *Nr3c1* expression is already detected at E10.5 and negative regulation of HPA axis components POMC and CRH occurs by E16.5 and is dependent upon GR [51]. Whether GCs are synthesized locally during skin development is not known, and is a subject for future investigation. We have evaluated and detected *Nr3c1* expression in mouse skin starting at E14.5, though it may be present at earlier stages. Interestingly, epidermal GR transcript and protein expression peaks at E16.5, the critical period for epidermal barrier acquisition, and decreases thereafter [52]. Importantly at E16.5, the skin levels of *Hsd11b1* and *Hsd11b2* are relatively low, compared to E18.5 when expression increases by more than 10- and 30-fold, respectively [38]. These data indicate that during barrier acquisition there is abundant receptor and a supply of active GCs in the skin (**Figure 2A**).

The first direct evidence that GCs regulate epidermal development was that antenatal exposure of rats to pharmacological doses of GCs accelerated permeability barrier acquisition, assessed functionally by measurements of transepidermal water loss (TEWL), and supported

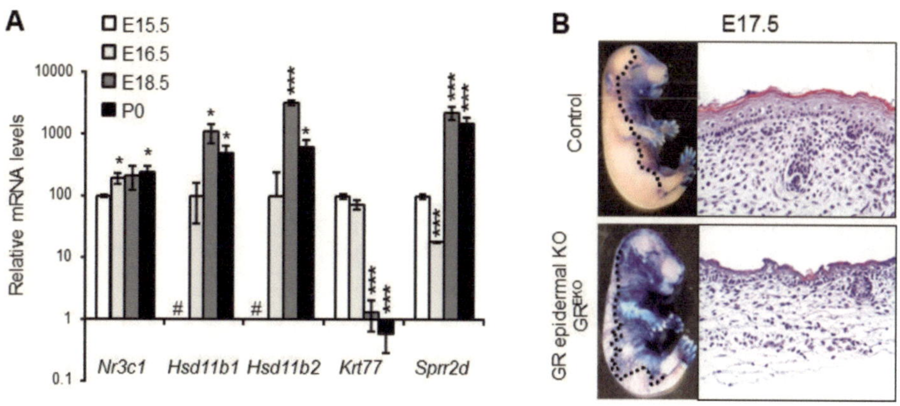

Figure 2. Relative gene expression of GR and enzymes modulating GC availability during embryonic skin development. (A) Relative mRNA levels of *Nr3c1*/GR, *Hsd11b1*, *Hsd11b2*, *Krt77*, and *Sprr2d* in embryonic (E15.5-E18.5) and newborn (P0) mouse skin. *Krt77* and *Sprr2d* are shown as markers of earlier and later epidermal development. #, not assessed. Statistically significant differences relative to E15.5 are indicated (n = 4 per age; *$p < 0.05$, **$p < 0.01$, ***$p < 0.001$). (B) Epidermal barrier is impaired in GR epidermal KO (GR^{EKO}) mice. Left: Toluidine blue staining of control and GR^{EKO} mice (E17.5) demonstrated delayed and altered epidermal barrier formation in GR^{EKO} mice. Dotted lines point to the dorsoventral and anteroposterior patterns of epidermal maturation. Right panel: hematoxylin and eosin-stained skin sections (E17.5) show immature thinner epidermis with abnormal differentiation of suprabasal layers in GR^{EKO} relative to control mice.

by ultrastructural data showing mature SC and lipid lamellar bodies and by increases in total SC lipid content [53]. Later, study of the GC-deficient $CRH^{-/-}$ mice revealed delays in epidermal development and barrier formation, with structurally immature lipid lamellar bodies and decreased lipid deposition and expression of SC proteins involucrin (IVL), loricrin (LOR), and filaggrin (FGN) [54]. These delays were evident at E17.5 when barrier should be fully functional and could be rescued by supplementation with GCs. Importantly, by birth $CRH^{-/-}$ mice had a structurally normal epidermis and SC, though its functionality was not assessed through TEWL measurements, suggesting a transient role of GCs in skin development. These initial studies were validated by whole mount dye exclusion assays with mouse embryos that had been exposed to pharmacological doses of GCs [55], which developed the permeability barrier earlier than controls (~½ day). A later study used gene profiling to determine that the critical time window for transcriptional responsiveness of skin to maternal GC-treatment is from E15.5 to E16.5 [55]. Genes found to be upregulated in GC-exposed skin included those encoding FGN and late cornified envelope proteins mapping to the epidermal differentiation complex, found on chromosome 3q in mice, which contains over 55 genes involved in keratinocyte terminal differentiation.

2.3. Mouse models for studying GR function in skin development

Normal skin development requires GCs and can be accelerated by maternal exposure to this hormone; however, it was necessary to evaluate specific gain- and loss-of-function GR mouse models to pinpoint the role of this NR in the skin (**Table 1**). Gain-of-function was assessed using

Mouse model	Description	Skin development	Adult skin homeostasis	Refs
$GR^{-/-}$	Global GR KO	Perinatal death Impaired epidermal barrier SC virtually absent Increased KC proliferation and apoptosis	--	[60]
$GR^{A458T/A458T}$	Global GR knock-in: transcription defective mutant A458T	Viable No skin phenotype	No skin phenotype	[60]
K5-GR	Overexpression of GR targeted to epidermis and derivatives (constitutive)	Viable (perinatal death at high doses of the transgene) Epidermal thinning; reduced HF number; dysplasic HF	Focal alopecia Delayed skin wound repair Reduced cutaneous inflammation and non melanoma skin cancer	[49,50, 84, 91]
K5-GR-TR	Overexpression of the transcription defective mutant GR (P493R, A494S) targeted to epidermis and derivatives (constitutive)	Viable No skin phenotype	Focal alopecia Delayed skin wound repair Partial reduction of cutaneous inflammation	[59, 91]
$K5\text{-}cre//GR^{loxP/loxP}$ (GR^{EKO})	GR epidermal KO (constitutive)	Impaired KC differentiation Increased KC proliferation and apoptosis Skin inflammation	Increased cutaneous inflammation, non-melanoma skin cancer, and melanocytic foci	[66, 83]
$K14\text{-}cre\text{-}ER^{T2}//GR^{loxP/loxP}$	GR epidermal KO (induced in adult)	--	Thickened skin Reduced KC differentiation Increased dermal inflammation	[52]

Table 1. Summary of the skin phenotypes of genetically modified mice with GR gain- and loss-of-function. *Abbreviations*: GR, glucocorticoid receptor; SC, stratum corneum; KC, keratinocyte; HF, hair follicle; K, keratin.

transgenic mice (K5-GR) that express GR under the control of the K5 promoter that is active in all stratified epithelia, including the basal layer of the epidermis, hair follicles, and sebaceous glands [49, 56]. The effects of high levels of GR overexpression during development were deleterious, causing perinatal lethality and lesions lacking epidermis and/or skin (**Table 1**). Transgenic mice overexpressing lower levels of GR survived to adulthood and were fertile, but also showed developmental abnormalities in morphogenesis of the epidermis and hair [49, 50]. During development and at birth, K5-GR epidermis is thinner than littermate controls, with drastic reductions in K5 expression [50], consistent with the antiproliferative effects of pharmacological GC treatments on keratinocytes [57, 58]. In addition, keratinocyte-specific GR overexpression caused accelerated epidermal differentiation as seen by increased LOR staining (our unpublished data). To evaluate the respective contributions of GR transactivation/transrepression in the phenotype of the K5-GR transgenic mice, another transgenic mouse model (K5-GR-TR) was generated with epidermal overexpression of GR carrying a double mutation (P493R and A494S) in the second half of the second zinc finger which impairs transcriptional activation but not transrepression of AP-1 and NF-κB [59]. The K5-GR-TR mice showed normal skin development without abnormalities in histology or in the expression of K5, K10, FGN, LOR, and IVL ([59] and our unpublished data) suggesting that dysregulation of epidermal markers in developing skin due to GR overexpression depends upon its ability to activate transcription (**Table 1**).

The analysis of the complete loss of function $GR^{-/-}$ mice revealed phenotypes complementary to the K5-GR model. These mice die upon birth due to defects in lung maturation, so analysis was unable to go beyond this stage [51]; however, a pronounced skin phenotype was observed in developing and newborn $GR^{-/-}$ animals featuring abnormal K5 expression in suprabasal layers, almost negligible levels of FGN, LOR, and IVL, and increased apoptosis. An increase in phosphorylated ERK was observed in $GR^{-/-}$ keratinocytes *in vivo* as well as *in vitro* where it was shown to contribute to the increased apoptosis [60]. Dye exclusion assays confirmed that the formation of the permeability barrier was delayed relative to control littermates (**Table 1**). The defective epidermal differentiation in the $GR^{-/-}$mice correlated with the altered expression of genes in the epidermal differentiation complex, with strong repression of members of the *Small proline repeat rich* family and *Corneodesmosin* (*Cdsn*) and upregulation of early differentiation genes such as the epithelial-specific gene transcripts E74-like factor 5 (*Elf5*) and keratin 77 (*Krt77*) [52]. In contrast, $GR^{A458T/A458T}$ knock-in mice were viable and fertile, and had normal histological appearance and expression of K5, K10, and LOR, suggesting that the transactivation function is not required for survival or skin development [60, 61]. It is worth noting that newborn mice with complete loss of the cytoplasmic chaperone p23 showed striking similarities in defects in skin development to those observed in $GR^{-/-}$ mice [60, 62], including defective keratinocyte differentiation, proliferation, and increased apoptosis. Indeed, GR-controlled target genes such as *Elf5*, *Krt77*, and *Cdsn* were also dysregulated in $p23^{-/-}$ skin and GR nuclear translocation upon GC treatment was defective in $p23^{-/-}$ cultured keratinocytes, indicating cell autonomous defects [62].

In order to study the effects of GR loss beyond birth as well as to assess cell-type specific contributions to skin development and homeostasis, knockouts were generated using $GR^{flox/flox}$ mice, which have the third exon, encoding part of the GR DNA binding domain, flanked by loxP sites reviewed in [63]. When crossed with transgenic mice that express the Cre recombinase, exon 3 is disrupted and the GR gene inactivated due to out-of-frame splicing and premature translational termination [64]. Newborn mice lacking mesenchymal expression of

GR (Dermo1-Cre//GR$^{flox/flox}$) had defects in dermal collagen and elastin production as well as histological abnormalities in the epidermis, which included rounded suprabasal keratinocytes and a dense SC [65]. These results indicate that dermal defects due to the lack of GR in fibroblasts impact the adjacent epidermis; however, as more detailed analysis was not performed, the exact mechanisms remain unclear. Mice lacking epidermal GR (K5-Cre//GR$^{flox/flox}$ or GR^{EKO}) showed defective skin development with delayed epidermal barrier formation, abnormal keratinocyte differentiation, hyperproliferation, and SC fragility [66] (**Table 1**). Toluidine blue dye exclusion assays revealed patchy disorganized barrier initiation sites and irregularities in barrier fronts, more evidence that GR regulates this process (**Figure 2B**). The mechanism is not entirely clear but may be related to negative regulation by GR of the kinase AKT and/or the AP-1 member Jun in the barrier front, as the activity of both must be spatiotemporally controlled for proper barrier acquisition [67]. Consistent with this phenotype was abnormal interfollicular K6 expression, decreased levels of LOR, FGN, and CDSN, and alterations in epidermal lipids. In addition to the barrier defects, GR^{EKO} mice had an inflammatory phenotype with increases in epidermal STAT3, AKT, and ERK activities and with dermal infiltrates containing macrophages and degranulated mast cells. Gene expression profiling data identified upregulation of *Elf5* and *Krt77* as well as the keratins *Krt6a*, *Krt6b*, and *Krt16*, which are induced in the context of hyperproliferative/inflammatory skin diseases, and many other genes such as *Tslp* and *S100a8/9* commonly induced in inflammatory skin diseases [66]. These data indicated that newborn GR^{EKO} mice suffer skin disease with features of atopic dermatitis and psoriasis. Loss-of-function mutation in epidermal barrier genes has been linked to atopic dermatitis and psoriasis, indicating that defects in keratinocyte terminal differentiation can be a predisposing factor for these diseases [11].

3. GR function in adult skin homeostasis

Research on the molecular mechanisms underlying GC actions has been put forward—at least to a great extent—because of the wide and efficacious use of GC-derived compounds for the treatment of chronic inflammatory diseases including those affecting skin. Contrary to the perinatal period, in which GCs accelerate skin barrier formation, GC treatment of adult animals perturbs permeability barrier homeostasis, suggesting unique roles for the GR in development and adulthood [29, 68].

3.1. Endogenous GCs affect skin integrity in aging and stress

Intrinsic or chronological skin aging affects nonexposed areas, and is mainly attributed to genetic factors and endocrine alterations. In contrast, extrinsic or pathological aging is principally due to repeated exposure to UV irradiation. In skin that is sun-exposed, both types of aging are superimposed [69]. Aging skin is characterized by gradual loss of the structural and functional characteristics of the tissue, which becomes more prone to damage, infections, and retarded wound healing, with consequent increases in the susceptibility of individuals to cutaneous disorders including those associated with inflammation and/or cancer [70]. Moreover, increased age in humans and mice correlates with abnormal skin barrier function with augmented TEWL and impaired mechanical properties partly due to a marked reduction in SC lipids resulting in decreased lipid layers, or mortar (see Section 2.1), between corneocytes [69]. Also, aged humans and mice typically have increased serum cytokines and markers of inflammation. Remarkably, correction of skin barrier defects in older mice by application of petrolatum or glycerol significantly reduced serum cytokines opening the attractive possibility that enhancing epidermal functions could ameliorate or prevent inflammation-associated disorders in elderly humans [71].

Another prominent feature of aged skin is atrophy, reduced epidermal, and dermal thickness caused by decreased keratinocyte proliferation, and profound alterations in extracellular matrix proteins of the dermis such as collagen, elastin, and proteoglycans, contributing to the formation of wrinkles and increased fragility [72]. GCs exert antiproliferative effects in skin inhibiting the proliferation of keratinocytes and fibroblasts. Transcriptomic analyses in human-cultured keratinocytes demonstrated that GCs regulate numerous genes participating in cytoskeletal rearrangements and ECM remodeling including *Actin* and *Krt6*, whose repression is consistent with the inhibition of keratinocyte migration and wound healing by GCs [23]. GCs also regulated numerous genes related to keratinocyte differentiation including *FLG* and *CDSN*. Additional studies in GC-treated adult mouse epidermis also identified numerous target genes related to cell cycle or DNA synthesis [73]. The induction of the stress-inducible mTOR inhibitor REDD1 contributed to the atrophogenic GC effects. As GCs elicited similar anti-inflammatory responses in control and *Redd1*$^{-/-}$ mice, the use of REDD1 inhibitors may have therapeutic implications [73].

The activity of HSD11B1 was increased in human and mouse skin samples from old relative to young subjects as well as in photodamaged versus nonexposed human skin biopsies, suggesting that local conversion of inactive to active GCs contributes to intrinsic and extrinsic aging of this tissue [74]. *Hsd11b1* KO mice were partially protected against age-induced skin damage showing increased collagen density as well as improved wound healing relative to controls [74]. Furthermore, topical treatment with a HSD11B1 inhibitor accelerated cutaneous wound healing in aged mice [74]. Altogether, HSD11B1 targeting appears as a promising pharmacological target to ameliorate cutaneous GC adverse side effects.

As endogenous GC production is also increased in stress conditions due to HPA reactivity, there is a link between pathologies with chronic-elevated GC levels and altered epidermal function. In particular, psychological stress is known to exacerbate features of skin diseases such as psoriasis and AD through increased GC production [75, 76]. In psychologically stressed mice, elevated GC levels inhibited epidermal lipid synthesis and downregulated the expression of antimicrobial peptides leading to decreased SC integrity and increased risk of infection [76, 77]. These defects could be reversed by reducing GC production through administration of an inhibitor of CRH or the GR antagonist RU486 and also by topical treatment with exogenous lipids [75, 76]. However, and paradoxically, it has also been reported that the stress-induced production of endogenous GCs exerted beneficial effects in cutaneous function in three different murine models of dermatoses, likely due to the anti-inflammatory effects of acute increases in endogenous GCs [78].

3.2. Skin alterations in adult transgenic mice with GR gain and loss of function

It is well known that GC treatment may cause marked epidermal thinning as well as retarded growth of hair follicles and hair loss, which may result in alopecia. Consistent with this, adult epidermis from K5-GR and K5-GR-TR mice showed pronounced epidermal hypoplasia with flattened keratinocytes and discontinuous K5 staining as well as reduction in the number of hair follicles (50 or 25% decrease, respectively) [49, 50, 59]. These data indicate that although GR-dependent transcriptional activation was partially impaired in K5-GR-TR mice, the overexpression of this GR mutant was sufficient to inhibit keratinocyte proliferation and alter hair follicle growth in adulthood but not during development [59]. Microarray studies in skin of K5-GR mice identified upregulation of a large subset of hair keratins, keratin-associated proteins, and downregulation of several *Hox* genes indicating a role of GR in hair follicle homeostasis through the control of keratin genes [79].

In addition, adult K5-GR mice exhibited abnormalities affecting other ectodermal derivatives, including exocrine glands such as the sweat glands, the ocular secretory Meibomian glands, and the preputial glands [49, 50]. In fact, the phenotype of K5-GR mice recapitulated the triad of clinical symptoms that defines the human syndrome hypohidrotic ectodermal dysplasia (hair, teeth, and exocrine glands) [50]. Although the exact mechanisms underlying these defects have not been characterized, the overexpression of GR impaired the expression and/or activity of NF-κB and p63 in several epithelia [50]. The fact that neither GR^{EKO} nor K5-GR-TR adult mice exhibited the ectodermal defects observed in adult K5-GR mice suggests that these abnormalities depend on elevated levels of transcriptionally competent GR.

The severe skin phenotype of newborn GR^{EKO} mice featuring barrier defects and inflammation resolved spontaneously around postnatal day 5. Adult mice showed only a mild phenotype of increased keratinocyte proliferation and patches of impaired epidermal differentiation indicating that barrier function is largely intact (**Table 1**). However, as in other models with impaired skin barrier development, epidermal GR loss resulted in increased susceptibility in adulthood to inflammatory triggers such as PMA, with elevated levels of K6 consistent with its upregulation in GR^{EKO} skin during development [66].

Also, the tamoxifen-inducible epidermal deletion of GR in adult mice (K14-Cre-ER^T//$GR^{flox/flox}$ mice) resulted in skin alterations with thickened epidermis, abnormal expression of K6 in the interfollicular epidermis, K10 localization restricted to the most suprabasal epidermal layer, reduced and patchy expression of LOR and CDSN, and the presence of dermal infiltrates [52]. After acute PMA treatment, K14-Cre-ER^T//$GR^{flox/flox}$ mice showed significantly increased keratinocyte proliferation and skin inflammation, with pronounced recruitment of polymorphonuclear cells [52]. An independently generated mouse model with tamoxifen-inducible GR epidermal deletion in adulthood also showed increased induction of TSLP, a key marker of atopic dermatitis, which could not be inhibited by GCs [22].

The posttranslational modifications of GR play an important role in the susceptibility to PMA-induced skin inflammation. It was recently demonstrated that GR sumoylation at K310 in mice (K293 in humans) is required for the formation of a repressing complex, which is involved in both GR-mediated IR nGRE gene repression and transrepression of NF-kB/AP-1-driven transcription. Mice harboring mutations that impaired GR sumoylation at this site showed more severe responses to PMA-induced skin inflammation, which could not efficiently suppressed by Dex [80, 81]. Experiments in mice with keratinocyte-specific inactivation of the components of the repressing complex NCoR1/SMRT or HDAC3 showed the lack of Dex-induced transrepression as tethering of the complex on DNA-bound NF-κB/AP1 was impaired [80, 81].

The recent finding that GR haploinsufficiency in mice and reduced GR expression in human biopsies correlates with increased incidence of tumor formation provides a causal role for this TF in tumorigenesis, reinforcing the denomination of GR as a tumor suppressor gene [82]. In fact, the analyses of mouse models with epidermal-specific GR overexpression or inactivation demonstrated that GR exerts tumor suppressor actions during skin carcinogenesis [80, 84]. GR^{EKO} mice subjected to the classical two-stage protocol—consisting in a single low-dose application of the mutagen 12-dimethylbenz(a) anthracene (DMBA) followed by repeated PMA treatments—exhibited earlier papilloma formation with higher incidence and multiplicity, as well as increased tumor size relative to controls [83]. Also, papillomas in GR^{EKO} mice displayed signs of early malignization, including delocalized expression of laminin A, dermal K5-positive cells, abnormal expression of K13, and focal loss of E-cadherin. Consistent with the keratinocyte atypia *in vivo*, cultured GR^{EKO} keratinocytes showed abnormal spindle-like morphology, loss of E-cadherin, and upregulation of smooth muscle actin and SNAIL, overall suggesting epithelial-mesenchymal transition [83]. Conversely, transgenic K5-GR//Ha-ras+ mice showed resistance to PMA-induced skin carcinogenesis with delayed onset of papilloma appearance, reduced tumor burden, and significant decrease of papilloma size (eightfold) relative to WT// Ha-ras+ controls [84]. Mechanistically, GR function in mouse skin tumorigenesis was mediated through negative interference with the NF-κB, AKT, and STAT3 pathways [83, 84]. It has also been postulated that the antitumor effects of GR in K5-GR//Ha-ras+ mice were exerted by decreasing the number of follicular stem cells as well as their proliferative potential, with associated changes in their transcriptional signature [85]. However, recent work suggests that increases in the local concentration of bioactive GCs can also exert tumor-promoting effects in solid tumors of epithelial origin [86]. Overall, additional studies are required to understand the apparently controversial data on the role of GC signaling on epithelial tumor development and progression.

Another major adverse effect associated with pharmacological GC treatment is delayed wound healing. Wound healing is a complex process comprising inflammatory, proliferative, and remodeling phases, which requires coordinated interactions among keratinocytes, immune cells, and fibroblasts to repair tissue damage and restore skin homeostasis [87]. Although inflammation is required for skin barrier restoration, alterations in levels or kinetics of expression of inflammatory mediators can be detrimental and result in chronic wounds or delayed healing [88]. Secretion of growth factors and cytokines including FGFs, EGF, IL-1, and IL-6 stimulates keratinocyte proliferation and migration, a process called re-epithelialization, normally accompanied by collagen deposition, formation of new granulation tissue, and wound contraction [87, 88].

Endogenous GC excess, for instance, in diabetic patients, results in chronic nonhealing wounds and often in lower limb amputations. GC-activated GR inhibits wound closure by blocking EGF-induced keratinocyte migration. The mechanism involves the formation of a repressor complex together with β-catenin to inhibit K6 and K16 expression at the wound edge [88]. Importantly, farnesyl pyrophosphate (FPP), an intermediate in the pathway of cholesterol biosynthesis, acts as GR agonist and also suppresses the *Krt6* gene and inhibits keratinocyte migration. The activation of GR by cortisol or FPP caused nuclear translocation of β-catenin, leading to induction of c-myc, a hallmark of chronic nonhealing wounds. This led to the proposal to use statins to restore epidermal homeostasis as targeting the cholesterol pathway interferes with the production of FPP and cortisol, resulting in substantial reduction of GR activation, and c-myc downregulation [89]. It has also been shown that GCs can also inhibit keratinocyte migration and wound healing by activation of nongenomic signaling pathways involving membrane GR regulation of phospholipase C/protein kinase C ultimately activating β-catenin and c-myc [90].

The use of K5-GR and K5-GR-TR mice allowed us to demonstrate that keratinocyte-targeted GR overexpression delayed skin wound healing [91]. This delay resulted from reducing the inflammatory response and decreasing keratinocyte migration *in vitro* and *in vivo*, consistent with the impaired skin healing observed with GC treatment [91]. While in K5-GR mice, cutaneous healing was delayed at days 4 and 8 after wounding, there was only a delay at day 4 in K5-GR-TR mice. These changes correlated with reduced K6 staining in both mouse models at day 4, which only persisted at day 8 in K5-GR, along with discontinuous K10 expression indicating an incomplete restoration of the epidermal barrier. These animals showed normal healing by day 8, concomitant with decreased repression of proinflammatory cytokines and growth factors relative to K5-GR mice. In wound healing experiments with both transgenic mouse models, keratinocyte proliferation was inhibited correlating with reduced ERK activity, *in vitro* and *in vivo*, and collagen deposition was reduced to a similar extent [91]. These data suggest that the early stages of wound closure are negatively regulated by GR independently of transcription, while GR transcriptional actions are necessary for delaying later stages of healing.

Finally, cutaneous production of GCs upon acute wounding is a major regulator of inflammatory responses as locally produced cortisol acts as a negative feedback to shut off the synthesis of the proinflammatory cytokine IL-1 [4, 25, 92]. The inhibition of skin-specific GC synthesis by using the 11β-hydroxylase inhibitor metyrapone or targeting HSD11B1 accelerated wound closure *in vivo*. These findings illustrate the relevance of local GCs to achieve a proper balance between pro- and anti-inflammatory signals upon injury, and thus modulate skin homeostasis. It has been also suggested that the interaction between cutaneous and systemic production of GCs has an impact on wound healing [31].

4. Conclusions

In vivo and *in vitro* studies, and in particular, the characterization of mouse models with gain- or loss-of-function of epidermal GR has highlighted a previously unrecognized role for this TF in skin development. Transcriptomic studies demonstrated a central role for GR in the regulation of epidermal genes, and specifically keratins with key roles in proliferation and differentiation including *Krt6* and *Krt77*. As in other models with impaired skin barrier development, epidermal GR loss resulted in increased susceptibility to inflammatory triggers in adulthood. The regulation of different biological processes and gene subsets by GR in skin was dependent on the developmental stage and physiological state, consistent with the context-specific actions of this TF. The identification of more specific downstream mediators of GC action with reduced adverse side effects remains a central objective in dermatological research. Also, the interactions between systemic and local HPA axes and GC production must be taken into consideration for developing novel strategies for treating cutaneous diseases.

Acknowledgements

We acknowledge to all former and present members of the lab who contributed to the work. We also acknowledge funding support from MINECO, Spanish Government (grants SAF2014-59474-R and SAF2017-88046-R). We thank NURCAMEIN (SAF2015-71878-REDT and SAF2017-90604-REDT) for support for dissemination.

Conflict of interest

The authors declare no conflict of interest.

Author details

Lisa M. Sevilla and Paloma Pérez*

*Address all correspondence to: pperez@ibv.csic.es

Instituto de Biomedicina de Valencia (IBV)-CSIC, Valencia, Spain

References

[1] Granner DK, Wang JC, Yamamoto KR. Regulatory actions of glucocorticoid hormones: From organisms to mechanisms. In: Wang JC, Harris C, editors. Advances in Experimental Medicine and Biology. Glucocorticoid Signaling From Molecules to Mice to Man. New York: Springer; 2015. pp. 3-31. DOI: 10.1007/978-1-4939-2895-8

[2] Kadmiel M, Cidlowski JA. Glucocorticoid receptor signaling in health and disease. Trends in Pharmacological Sciences. 2013;**34**:518-530. DOI: 10.1016/j.tips.2013.07.003

[3] Taves MD, Gomez-Sanchez CE, Soma KK. Extra-adrenal glucocorticoids and mineralocorticoids: Evidence for local synthesis, regulation, and function. American Journal of Physiology. Endocrinology and Metabolism. 2011;**301**:E11-E24. DOI: 10.1152/ajpendo.00100.2011

[4] Slominski AT, Manna PR, Tuckey RC. Cutaneous glucocorticosteroidogenesis: Securing local homeostasis and the skin integrity. Experimental Dermatology. 2014;**23**:369-374. DOI: 10.1111/exd.12376

[5] Chapman K, Holmes M, Seckl J. 11β-hydroxysteroid dehydrogenases: Intracellular gatekeepers of tissue glucocorticoid action. Physiological Reviews. 2013;**93**(3):1139-1206. DOI: 10.1152/physrev.00020.2012

[6] Weikum ER, Knuesel MT, Ortlund EA, Yamamoto KR. Glucocorticoid receptor control of transcription: Precision and plasticity via allostery. Nature Reviews. Molecular Cell Biology. 2017;**18**(3):159-174. DOI: 10.1038/nrm.2016.152

[7] Sacta MA, Chinenov Y, Rogatsky I. Glucocorticoid signaling: An update from a genomic perspective. Annual Review of Physiology. 2016;**78**:155-180. DOI: 10.1146/annurev-physiol-021115-105323

[8] Vandewalle J, Luypaert A, De Bosscher K, Libert C. Therapeutic mechanisms of glucocorticoids. Trends in Endocrinology and Metabolism. 2018;**29**:42-54. DOI: 10.1016/j.tem.2017.10.010

[9] Desmet SJ, De Bosscher K. Glucocorticoid receptors: Finding the middle ground. The Journal of Clinical Investigation. 2017;**127**(4):1136-1145. DOI: 10.1172/JCI88886

[10] Weidinger S, Novak N. Atopic dermatitis. Lancet. 2016;**387**(10023):1109-1122. DOI: 10.1016/S0140-6736(15)00149-X

[11] Boehncke WH, Schön MP. Psoriasis. Lancet. 2015;**386**(9997):983-994. DOI: 10.1016/S0140-6736(14)61909-7

[12] Ramot Y, Paus R, Tiede S, Zlotogorski A. Endocrine controls of keratin expression. BioEssays. 2009;**31**(4):389-399. DOI: 10.1002/bies.200800121

[13] Presman DM, Ogara MF, Stortz M, Alvarez LD, Pooley JR, Schiltz RL, Grøntved L, Johnson TA, Mittelstadt PR, Ashwell JD, Ganesan S, Burton G, Levi V, Hager GL, Pecci A. Live cell imaging unveils multiple domain requirements for in vivo dimerization of the glucocorticoid receptor. PLoS Biology. 2014;**12**(3):e1001813. DOI: 10.1371/journal.pbio.1001813

[14] Hägg PM, Hurskainen T, Palatsi R, Ilves M, Oikarinen A. Increased expression of glucocorticoid receptor beta in lymphocytes of patients with severe atopic dermatitis unresponsive to topical corticosteroid. The British Journal of Dermatology. 2010;**162**(2):318-324. DOI: 10.1111/j.1365-2133.2009.09518.x

[15] Kubin ME, Hägg PM, Kokkonen N, Väyrynen JP, Haapasaari KM, Moilanen J, Kallioinen M, Hurskainen T, Tasanen K. Glucocorticoid receptors GRα and GRβ are expressed in inflammatory dermatoses. European Journal of Dermatology. 2016;**26**(1):21-27. DOI: 10.1684/ejd.2015.2691

[16] So AY, Chaivorapol C, Bolton EC, Li H, Yamamoto KR. Determinants of cell- and gene-specific transcriptional regulation by the glucocorticoid receptor. PLoS Genetics. 2007;**3**(6):e94

[17] Reddy TE, Pauli F, Sprouse RO, Neff NF, Newberry KM, Garabedian MJ, Myers RM. Genomic determination of the glucocorticoid response reveals unexpected mechanisms of gene regulation. Genome Research. 2009;**19**(12):2163-2171. DOI: 10.1101/gr.097022.109

[18] Meijsing SH, Pufall MA, So AY, Bates DL, Chen L, Yamamoto KR. DNA binding site sequence directs glucocorticoid receptor structure and activity. Science. 2009;**324**(5925):407-410. DOI: 10.1126/science.1164265

[19] Schiller BJ, Chodankar R, Watson LC, Stallcup MR, Yamamoto KR. Glucocorticoid receptor binds half sites as a monomer and regulates specific target genes. Genome Biology. 2014;**15**(7):418. DOI: 10.1186/s13059-014-0418-y

[20] Radoja N, Komine M, Jho SH, Blumenberg M, Tomic-Canic M. Novel mechanism of steroid action in skin through glucocorticoid receptor monomers. Molecular and Cellular Biology. 2000;**20**(12):4328-4339

[21] Jho SH, Radoja N, Im MJ, Tomic-Canic M. Negative response elements in keratin genes mediate transcriptional repression and the cross-talk among nuclear receptors. The Journal of Biological Chemistry. 2001;**276**(49):45914-45920

[22] Surjit M, Ganti KP, Mukherji A, Ye T, Hua G, Metzger D, Li M, Chambon P. Widespread negative response elements mediate direct repression by agonist-liganded glucocorticoid receptor. Cell. 2011;**145**(2):224-241. DOI: 10.1016/j.cell.2011.03.027

[23] Petta I, Dejager L, Ballegeer M, Lievens S, Tavernier J, De Bosscher K, Libert C. The interactome of the glucocorticoid receptor and its influence on the actions of glucocorticoids in combatting inflammatory and infectious diseases. Microbiology and Molecular Biology Reviews. 2016;**80**:495-522. DOI: 10.1128/MMBR.00064-15

[24] Sevilla LM, Latorre V, Carceller E, Boix J, Vodák D, Mills IG, Pérez P. Glucocorticoid receptor and Klf4 co-regulate anti-inflammatory genes in keratinocytes. Molecular and Cellular Endocrinology. 2015;**412**:281-289. DOI: 10.1016/j.mce.2015.05.015

[25] Stojadinovic O, Lee B, Vouthounis C, Vukelic S, Pastar I, Blumenberg M, et al. Novel genomic effects of glucocorticoids in epidermal keratinocytes: Inhibition of apoptosis, interferon-gamma pathway, and wound healing along with promotion of terminal differentiation. The Journal of Biological Chemistry. 2007;**282**:4021-4034

[26] Leis H, Page A, Ramírez A, Bravo A, Segrelles C, Paramio J, Barettino D, Jorcano JL, Pérez P. Glucocorticoid receptor counteracts tumorigenic activity of Akt in skin through interference with the phosphatidylinositol 3-kinase signaling pathway. Molecular Endocrinology. 2004;**18**(2):303-311

[27] Nicolaides NC, Lamprokostopoulou A, Sertedaki A, Charmandari E. Recent advances in the molecular mechanisms causing primary generalized glucocorticoid resistance. Hormones. 2016;**15**:23-34. DOI: 10.14310/horm.2002.1660

[28] Schoepe S, Schäcke H, May E, Asadullah K. Glucocorticoid therapy-induced skin atrophy. Experimental Dermatology. 2006;**15**:406-420

[29] Terao M, Katayama I. Local cortisol/corticosterone activation in skin physiology and pathology. Journal of Dermatological Science. 2016;**84**:11-16. DOI: 10.1016/j.jdermsci.2016.06.014

[30] Nikolakis G, Stratakis CA, Kanaki T, Slominski A, Zouboulis CC. Skin steroidogenesis in health and disease. Reviews in Endocrine & Metabolic Disorders. 2016;**17**(3):247-258

[31] Skobowiat C, Slominski AT. UVB activates hypothalamic-pituitary-adrenal axis in C57BL/6 mice. The Journal of Investigative Dermatology. 2015;**135**(6):1638-1648. DOI: 10.1038/jid.2014.450

[32] Sarkar MK, Kaplan N, Tsoi LC, Xing X, Liang Y, Swindell WR, et al. Endogenous glucocorticoid deficiency in psoriasis promotes inflammation and abnormal differentiation. The Journal of Investigative Dermatology. 2017;**137**:1474-1483

[33] Hannen R, Udeh-Momoh C, Upton J, Wright M, Michael A, Gulati A, Rajpopat S, Clayton N, Halsall D, Burrin J, Flower R, Sevilla L, Latorre V, Frame J, Lightman S, Pérez P, Philpott M. Dysfunctional skin-derived glucocorticoid synthesis is a pathogenic mechanism of psoriasis. The Journal of Investigative Dermatology. 2017;**137**(8):1630-1637. DOI: 10.1016/j.jid.2017.02.984

[34] Martinerie L, Munier M, Menuet DL, Meduri G, Viengchareun S, Lombes M. The mineralocorticoid signaling pathway throughout development: Expression, regulation and pathophysiological implications. Biochimie. 2013;**95**:148-157. DOI: 10.1016/j.biochi.2012.09.030

[35] Gomez-Sanchez E, Gomez-Sanchez CE. The multifaceted mineralocorticoid receptor. Comprehensive Physiology. 2014;**4**:965-994. DOI: 10.1002/cphy.c130044

[36] Tiganescu A, Walker EA, Hardy RS, Mayes AE, Stewart PM. Localization, age- and site-dependent expression, and regulation of 11β-hydroxysteroid dehydrogenase type 1 in skin. The Journal of Investigative Dermatology. 2011;**131**(1):30-36. DOI: 10.1038/jid.2010.257

[37] Cirillo N, Prime SS. Keratinocytes synthesize and activate cortisol. Journal of Cellular Biochemistry. 2011;**112**(6):1499-1505. DOI: 10.1002/jcb.23081

[38] Boix J, Carceller E, Sevilla LM, Marcos-Garcés V, Pérez P. The mineralocorticoid receptor plays a transient role in mouse skin development. Experimental Dermatology. 2016;**25**(1):69-71. DOI: 10.1111/exd.12822

[39] Boix J, Bigas J, Sevilla LM, Iacobone M, Citton M, Torresan F, Caroccia B, Rossi GP, Pérez P. Primary aldosteronism patients show skin alterations and abnormal activation of glucocorticoid receptor in keratinocytes. Scientific Reports. 2017;**7**(1):15806. DOI: 10.1038/s41598-017-16216-5

[40] Jaisser F, Farman N. Emerging roles of the mineralocorticoid receptor in pathology: Toward new paradigms in clinical pharmacology. Pharmacological Reviews. 2016;**68**:49-75. DOI: 10.1124/pr.115.011106

[41] Liu S, Zhang H, Duan E. Epidermal development in mammals: Key regulators, signals from beneath, and stem cells. International Journal of Molecular Sciences. 2013;**14**(6):10869-10895. DOI: 10.3390/ijms140610869

[42] Gonzales KAU, Fuchs E. Skin and its regenerative powers: An alliance between stem cells and their niche. Developmental Cell. 2017;**43**:387-401. DOI: 10.1016/j.devcel.2017.10.001

[43] Elias PM. Stratum corneum defensive functions: An integrated view. The Journal of Investigative Dermatology. 2005;**125**:183-200

[44] Hardman MJ, Sisi P, Banbury DN, Byrne C. Patterned acquisition of skin barrier function during development. Development. 1998;**125**(8):1541-1552

[45] Elias PM. Therapeutic implications of a barrier-based pathogenesis of atopic dermatitis. Annals of Dermatology. 2010;**22**(3):245-254. DOI: 10.5021/ad.2010.22.3.245

[46] Watt FM. Mammalian skin cell biology: At the interface between laboratory and clinic. Science. 2014;**346**:937-940. DOI: 10.1126/science.1253734

[47] Soares E, Zhou H. Master regulatory role of p63 in epidermal development and disease. Cellular and Molecular Life Sciences. 2018;**75**(7):1179-1190. DOI: 10.1007/s00018-017-2701-z

[48] Klein RH, Lin Z, Hopkin AS, Gordon W, Tsoi LC, Liang Y, Gudjonsson JE, Andersen B. GRHL3 binding and enhancers rearrange as epidermal keratinocytes transition between functional states. PLoS Genetics. 2017;**13**(4):e1006745. DOI: 10.1371/journal.pgen.1006745

[49] Pérez P, Page A, Bravo A, Del Río M, Giménez-Conti I, Budunova I, Slaga TJ, Jorcano JL. Altered skin development and impaired proliferative and inflammatory responses in transgenic mice overexpressing the glucocorticoid receptor. The FASEB Journal. 2001;**15**: 2030-2032

[50] Cascallana JL, Bravo A, Donet E, Leis H, Lara MF, Paramio JM, Jorcano JL, Pérez P. Ectoderm-targeted overexpression of the glucocorticoid receptor induces hypohidrotic ectodermal dysplasia. Endocrinology. 2005;**146**:2629-2638

[51] Cole TJ, Blendy AP, Monaghan K, Schmid W, Aguzzi A, Fantuzzi G, et al. Targeted disruption of the glucocorticoid receptor gene blocks adrenergic chromaffin cell development and severely retards lung maturation. Genes & Development. 1995;**9**:1608-1621

[52] Sevilla LM, Bayo P, Latorre V, Sanchis A, Pérez P. Glucocorticoid receptor regulates overlapping and differential gene subsets in developing and adult skin. Molecular Endocrinology. 2010;**24**(11):2166-2178. DOI: 10.1210/me.2010-0183

[53] Aszterbaum M, Feingold KR, Menon GK, Williams ML. Glucocorticoids accelerate fetal maturation of the epidermal permeability barrier in the rat. The Journal of Clinical Investigation. 1993;**91**(6):2703-2708

[54] Hanley K, Feingold KR, Kömüves LG, Elias PM, Muglia LJ, Majzoub JA, Williams ML. Glucocorticoid deficiency delays stratum corneum maturation in the fetal mouse. The Journal of Investigative Dermatology. 1998;**111**(3):440-444

[55] Segre JA. Epidermal barrier formation and recovery in skin disorders. The Journal of Clinical Investigation. 2006;**116**:1150-1158

[56] Ramírez A, Bravo A, Jorcano JL, Vidal M. Sequences 5′ of the bovine keratin 5 gene direct tissue- and cell-type-specific expression of a lacZ gene in the adult and during development. Differentiation. 1994;**58**(1):53-64

[57] Schäcke H, Döcke WD, Asadullah K. Mechanisms involved in the side effects of glucocorticoids. Pharmacology & Therapeutics. 2002;**96**(1):23-43

[58] Budunova IV, Carbajal S, Kang H, Viaje A, Slaga TJ. Altered glucocorticoid receptor expression and function during mouse skin carcinogenesis. Molecular Carcinogenesis. 1997;**18**(3):177-185

[59] Donet E, Bosch P, Sanchis A, Bayo P, Ramírez A, Cascallana JL, Bravo A, Pérez P. Transre pression function of the glucocorticoid receptor regulates eyelid development and keratinocyte proliferation but is not sufficient to prevent skin chronic inflammation. Molecular Endocrinology. 2008;**22**:799-812

[60] Bayo P, Sanchis A, Bravo A, Cascallana JL, Buder K, Tuckermann J, Schütz G, Pérez P. Gluco corticoid receptor is required for skin barrier competence. Endocrinology. 2008;**149**: 1377-1388

[61] Reichardt HM, Kaestner KH, Tuckermann J, Kretz O, Wessely O, Bock R, Gass P, Schmid W, Herrlich P, Angel P, Schütz G. DNA binding of the glucocorticoid receptor is not essential for survival. Cell. 1998;**93**(4):531-541

[62] Madon-Simon M, Grad I, Bayo P, Pérez P, Picard D. Defective glucocorticoid receptor signaling and keratinocyte-autonomous defects contribute to skin phenotype of mouse embryos lacking the Hsp90 co-chaperone p23. PLoS One. 2017;**12**(6):e0180035. DOI: 10. 1371/journal.pone.0180035

[63] Whirledge S, DeFranco DB. Glucocorticoid signaling in health and disease: Insights from tissue-specific GR knockout mice. Endocrinology. 2018;**159**:46-64. DOI: 10.1210/en. 2017-00728

[64] Tronche F, Kellendonk C, Kretz O, Gass P, Anlag K, Orban PC, Bock R, Klein R, Schütz G. Disruption of the glucocorticoid receptor gene in the nervous system results in reduced anxiety. Nature Genetics. 1999;**23**(1):99-103

[65] Li A, Hardy R, Stoner S, Tuckermann J, Seibel M, Zhou H. Deletion of mesenchymal glucocorticoid receptor attenuates embryonic lung development and abdominal wall closure. PLoS One. 2013;**8**(5):e63578. DOI: 10.1371/journal.pone.0063578

[66] Sevilla LM, Latorre V, Sanchis A, Pérez P. Epidermal inactivation of the glucocorticoid receptor triggers skin barrier defects and cutaneous inflammation. The Journal of Investigative Dermatology. 2013;**133**(2):361-370. DOI: 10.1038/jid.2012.281

[67] O'Shaughnessy RF, Welti JC, Sully K, Byrne C. Akt-dependent Pp2a activity is required for epidermal barrier formation during late embryonic development. Development. 2009; **136**(20):3423-3431. DOI: 10.1242/dev.037010

[68] Kao JS, Fluhr JW, Man MQ, Fowler AJ, Hachem JP, Crumrine D, Ahn SK, Brown BE, Elias PM, Feingold KR. Short-termglucocorticoid treatment compromises both permeability barrier homeostasis and stratum corneum integrity: Inhibition of epidermal lipid synthesis accounts for functional abnormalities. The Journal of Investigative Dermatology. 2003;**120**:456-464

[69] Ghadially R, Brown BE, Sequeira-Martin SM, Feingold KR, Elias PM. The aged epidermal permeability barrier. Structural, functional, and lipid biochemical abnormalities in humans and a senescent murine model. The Journal of Clinical Investigation. 1995;**95**(5): 2281-2290

[70] Hwang KA, Yi BR, Choi KC. Molecular mechanisms and in vivo mouse models of skin aging associated with dermal matrix alterations. Laboratory Animal Research. 2011;**27**(1): 1-8. DOI: 10.5625/lar.2011.27.1.1

[71] Hu L, Mauro TM, Dang E, Man G, Zhang J, Lee D, Wang G, Feingold KR, Elias PM, Man MQ. Epidermal dysfunction leads to an age-associated increase in levels of serum inflammatory cytokines. The Journal of Investigative Dermatology. 2017;**137**(6):1277-1285. DOI: 10.1016/j.jid.2017.01.007

[72] Cole MA, Quan T, Voorhees JJ, Fisher GJ. Extracellular matrix regulation of fibroblast function: Redefining our perspective on skin aging. Journal of Cell Communication and Signaling. 2018;**12**(1):35-43. DOI: 10.1007/s12079-018-0459-1

[73] Baida G, Bhalla P, Kirsanov K, Lesovaya E, Yakubovskaya M, Yuen K, Guo S, Lavker RM, Readhead B, Dudley JT, Budunova I. REDD1 functions at the crossroads between the therapeutic and adverse effects of topical glucocorticoids. EMBO Molecular Medicine. 2015;**7**(1):42-58. DOI: 10.15252/emmm.201404601

[74] Tiganescu A, Tahrani AA, Morgan SA, Otranto M, Desmoulière A, Abrahams L, Hassan-Smith Z, Walker EA, Rabbitt EH, Cooper MS, Amrein K, Lavery GG, Stewart PM. 11β-Hydroxysteroid dehydrogenase blockade prevents age-induced skin structure and function defects. The Journal of Clinical Investigation. 2013;**123**:3051-3060. DOI: 10.1172/JCI 64162

[75] Denda M, Tsuchiya T, Elias PM, Feingold KR. Stress alters cutaneous permeability barrier homeostasis. American Journal of Physiology. Regulatory, Integrative and Comparative Physiology. 2000;**278**:R367-R372

[76] Choi EH, Demerjian M, Crumrine D, Brown BE, Mauro T, Elias PM, Feingold KR. Glucocorticoid blockade reverses psychological stress-induced abnormalities in epidermal structure and function. American Journal of Physiology. Regulatory, Integrative and Comparative Physiology. 2006;**291**(6):R1657-R1662

[77] Aberg KM, Radek KA, Choi EH, Kim DK, Demerjian M, Hupe M, Kerbleski J, Gallo RL, Ganz T, Mauro T, Feingold KR, Elias PM. Psychological stress downregulates epidermal antimicrobial peptide expression and increases severity of cutaneous infections in mice. The Journal of Clinical Investigation. 2007;**117**:3339-3349

[78] Lin TK, Man MQ, Santiago JL, Scharschmidt TC, Hupe M, Martin-Ezquerra G, Youm JK, Zhai Y, Trullas C, Feingold KR, Elias PM. Paradoxical benefits of psychological stress in inflammatory dermatoses models are glucocorticoid mediated. The Journal of Investigative Dermatology. 2014;**134**(12):2890-2897. DOI: 10.1038/jid.2014.265

[79] Donet E, Bayo P, Calvo E, Labrie F, Pérez P. Identification of novel glucocorticoid receptor-regulated genes involved in epidermal homeostasis and hair follicle differentiation. The Journal of Steroid Biochemistry and Molecular Biology. 2008;**108**(1-2):8-16

[80] Hua G, Paulen L, Chambon P. GR SUMOylation and formation of an SUMO-SMRT/NCoR1-HDAC3 repressing complex is mandatory for GC-induced IR nGRE-mediated transrepression. Proceedings of the National Academy of Sciences of the United States of America. 2016;**113**(5):E626-E634. DOI: 10.1073/pnas.1522821113

[81] Hua G, Ganti KP, Chambon P. Glucocorticoid-induced tethered transrepression requires SUMOylation of GR and formation of a SUMO-SMRT/NCoR1-HDAC3 repressing complex. Proceedings of the National Academy of Sciences of the United States of America. 2016;**113**(5):E635-E643. DOI: 10.1073/pnas.1522826113

[82] Matthews LC, Berry AA, Morgan DJ, Poolman TM, Bauer K, Kramer F, Spiller DG, Richardson RV, Chapman KE, Farrow SN, Norman MR, Williamson AJ, Whetton AD, Taylor SS, Tuckermann JP, White MR, Ray DW. Glucocorticoid receptor regulates accurate chromosome segregation and is associated with malignancy. Proceedings of the National Academy of Sciences of the United States of America. 2015;**112**(17):5479-5484. DOI: 10.1073/pnas.1411356112

[83] Latorre V, Sevilla LM, Sanchis A, Pérez P. Selective ablation of glucocorticoid receptor in mouse keratinocytes increases susceptibility to skin tumorigenesis. The Journal of Investigative Dermatology. 2013;**133**:2771-2779. DOI: 10.1038/jid.2013.255

[84] Budunova IV, Kowalczyk D, Pérez P, Yao YJ, Jorcano JL, Slaga TJ. Glucocorticoid receptor functions as a potent suppressor of mouse skin carcinogenesis. Oncogene. 2003;**22**: 3279-3287

[85] Chebotaev D, Yemelyanov A, Zhu L, Lavker RM, Budunova I. The tumor suppressor effect of the glucocorticoid receptor in skin is mediated via its effect on follicular stem cells. Oncogene. 2007;**26**:3060-3068

[86] Azher S, Azami O, Amato C, McCullough M, Celentano A, Cirillo N. The non-conventional effects of glucocorticoids in cancer. Journal of Cellular Physiology. 2016;**231**(11):2368-2373. DOI: 10.1002/jcp.25408

[87] Werner S, Grose R. Regulation of wound healing by growth factors and cytokines. Physiological Reviews. 2003;**83**(3):835-870

[88] Pastar I, Stojadinovic O, Yin NC, Ramirez H, Nusbaum AG, Sawaya A, Patel SB, Khalid L, Isseroff RR, Tomic-Canic M. Epithelialization in wound healing: A comprehensive review. Advances in Wound Care (New Rochelle). 2014;**3**(7):445-464. DOI: 10.1089/wound.2013.0473

[89] Sawaya AP, Pastar I, Stojadinovic O, Lazovic S, Davis SC, Gil J, Kirsner RS, Tomic-Canic M. Topical mevastatin promotes wound healing by inhibiting the transcription factor c-Myc via the glucocorticoid receptor and the long non-coding RNA Gas5. The Journal of Biological Chemistry. 2018;**293**(4):1439-1449. DOI: 10.1074/jbc.M117.811240

[90] Jozic I, Vukelic S, Stojadinovic O, Liang L, Ramirez HA, Pastar I, Tomic Canic M. Stress signals, mediated by membranous glucocorticoid receptor, activate PLC/PKC/GSK-3β/β-catenin pathway to inhibit wound closure. The Journal of Investigative Dermatology. 2017;**137**(5):1144-1154. DOI: 10.1016/j.jid.2016.11.036

[91] Sanchis A, Alba L, Latorre V, Sevilla LM, Pérez P. Keratinocyte-targeted overexpression of the glucocorticoid receptor delays cutaneous wound healing. PLoS One. 2012;**7**(1): e29701. DOI: 10.1371/journal.pone.0029701

[92] Vukelic S, Stojadinovic O, Pastar I, Rabach M, Krzyzanowska A, Lebrun E, Davis SC, Resnik S, Brem H, Tomic-Canic M. Cortisol synthesis in epidermis is induced by IL-1 and tissue injury. The Journal of Biological Chemistry. 2011;**286**(12):10265-10275. DOI: 10.1074/jbc.M110.188268

Chapter 4

Neuroendocrine Controls of Keratin Expression in Human Skin

Yuval Ramot and Ralf Paus

Additional information is available at the end of the chapter

http://dx.doi.org/10.5772/intechopen.80406

Abstract

The human skin serves as a source for a large number of neurohormones and neuropeptides, which affect skin biology on multiple different levels. Intriguingly, this includes the control of keratin expression by neurohormones such as thyrotropin-releasing hormone, thyrotropin, opioids, prolactin, and cannabinoid receptor 1-ligands. While this neuroendocrine regulation of human keratin biology *in situ* is likely to be involved in the maintenance of skin and hair follicle homeostasis and may participate in skin pathology, this regulation remains to be appreciated and explored by mainstream keratin research. Here, we review recent progress in this frontier of neuroendocrine and keratin skin research, define the many open questions in the field, and elaborate how neurohormones may be harnessed to treat selected genodermatoses and other skin disorders accompanied by abnormal keratin expression.

Keywords: keratins, neuroendocrinology, hair, skin, dermatology

1. Introduction

Keratins are the major constituents of the epidermis and skin appendages, which by forming an intracellular structural network provide cellular stability and resilience to the tissue [1]. Furthermore, they exert a surprisingly wide and complex range of additional functions in the skin, including regulating epithelial differentiation and proliferation, migration and wound healing, carcinogenesis and apoptosis, and immunomodulation [2–5]. Taking into consideration the key roles keratins play in the skin, it is of utmost importance to understand and dissect the mediators that affect their expression. One of the key mediators of skin function is the endocrine system, which is also expressed and active in the skin itself.

An important pathway by which the endocrine system controls skin function is by changing keratin expression, and these effects have been described in detail previously [6].

Throughout the last decade, it became clear that the skin reacts and generates not only steroid hormones, but also a large array of neuroendocrine mediators [2, 3, 7–10]. The skin has even formed a hypothalamic-pituitary-adrenal (HPA) neuroendocrine signaling axis, equivalent to the central axis [10–12], and a semi-equivalent hypothalamic-pituitary-thyroid (HPT) axis [13–16]. These neuroendocrine mediators take part in the regulation of many different processes and functions of the skin, both in normal healthy skin and in disease states. These include, for example, regulation of stress response [10, 17], hair follicle (HF) growth [18–22], pigmentation of the skin and HF [18–21, 23, 24], sebaceous gland function [10, 12], proliferation and apoptosis of keratinocytes [9, 10, 25], and mitochondrial activity [16, 26, 27]. They are also involved in controlling the immune privilege of the HF epithelium and the immune response of the skin [24, 28].

Taking into consideration the fact that keratins constitute up to 85% of the cell mass of a terminally differentiated keratinocyte and have such important roles not only in keratinocyte, sebocyte, and trichocyte biology, but also for overall skin physiology [29–31] and the fact that the vast majority of neuroendocrine mediators is expressed in the skin epithelium [11, 12, 17], it is reasonable to ask whether some of the functions exerted by neurohormones in the skin are actually mediated by changing keratin expression. Indeed, in recent years, several studies have demonstrated that keratin expression in human skin and HFs is manipulated by neurohormones and underlies previously ignored, important neuroendocrine controls that invite therapeutic targeting.

In this chapter, we systematically explore the effects of neuroendocrine mediators on keratin expression and connect these changes to physiologically relevant functions of the skin and HFs. We also dissect the ways by which such keratin changes might be harnessed to alleviate different skin conditions.

2. The hypothalamic-pituitary-thyroid axis in the skin and its effects on keratin expression

The fact that skin and HFs are prominent targets for the thyroid hormones, triiodothyronine and thyroxine, is well established [15, 16]. These thyroid hormones also promote cutaneous wound healing [32, 33]. Furthermore, patients suffering from thyroid disorders manifest with significant hair and skin phenotypes [15]. It is possible that some of these changes are due to an effect of thyroid hormones on keratin expression. For example, T3 increases K6, K16, and K17 gene expression in human keratinocytes in culture, keratins that are known to be upregulated during the wound healing process [34], and mice with hypothyroidism have reduced K6 expression [34]. In addition, T3 and T4 stimulate K6 expression and decrease K14 expression in cultured human HFs [15].

However, thyroid hormones can themselves change the production of neurohormones such as prolactin and thyroid-stimulating hormone (TSH, thyrotropin), also in the skin [13, 35].

Indeed, in recent years, it has become evident that the skin expresses receptors for the thyroid hormones and for TSH and thyrotropin-releasing hormone (TRH) [13, 18, 23, 26]. It has also been observed that, just as in the central HPT axis, thyroid hormones decrease intraepidermal TSH expression, while TRH stimulates it in human skin, therefore suggesting that an elementary functional HPT axis also exists in the human skin [36].

Thyrotropin-releasing hormone is expressed by the human HF and can be found in the outer root sheath (ORS). The TRH receptor (TRH-R), on the other hand, is expressed in the inner root sheath (IRS) of the HF [23]. TRH can affect keratin expression: it has been found to upregulate the expression of the hair keratins K31 and K32, while it downregulates the expression of the hair keratins K85 and K86 at the protein level [37]. TRH also has profound effect on the keratins expressed by the ORS in the HF, leading to reduced expression of K6, K14, and K17 [23, 37]. The above-listed keratins have been confirmed to be regulated by TRH at the protein level in the HF, but it should be noted that additional keratins and keratin-associated proteins (KAPs) may be affected by TRH according to microarray results obtained with organ-cultured human HFs [37]. However, further experiments are required to confirm regulation of these keratins and KAPs by TRH. Another important open question is to which extent the TRH-induced changes in keratin expression observed in the HF underlie the complex functional changes exerted by TRH in the HF [2, 16], namely, the stimulation of hair shaft production by TRH [23].

In contrast to the ORS of the HF, TRH stimulated K6, K14, and K17 expression in the epidermis, sweat glands, and sebaceous glands in human skin *ex vivo* at the protein and mRNA levels [37]. The same promoting effect of TRH on human K6 expression was also evident in frog skin *in vitro* [25], and this stimulating effect was suggested to accompany the promotion of wound healing in the frog skin [25]. This suggests that the keratin regulatory effects of TRH are highly conserved in vertebrate skin and underscores the functional importance of this neuroendocrine control of keratin biology. This makes it even more surprising that mainstream keratin research continues to largely ignore this evolutionarily conserved control mechanism, which must have provided significant species survival advantages to have been maintained from frogs to humans. Interestingly, previous studies have found that TRH can also stimulate mitochondrial activity in human epidermis and scalp HFs [26]. This invites the intriguing question whether the part of this TRH-induced increased mitochondrial activity, and thus energy metabolism is actually recruited to promote and support the energy intensive synthesis of selected keratins.

Thyroid-stimulating hormone is another key neurohormone involved in the regulation of keratin expression in human skin. TSH is expressed in the epidermis, and the gene encoding its receptor reportedly is also transcribed in the epidermis [14], while TSH-R protein is most prominently, if not exclusively, found in the skin mesenchyme, including the dermal sheath of human scalp HFs [18]. However, there is still a debate on the exact location of the TSH-R protein [13, 38]. In whole skin organ cultures, TSH stimulated the expression of K5 and K14, the two prototypic keratins that are expressed in the basal layer of the epidermis, connect to the hemidesmosomes in the basal side of the keratinocytes and are critical for keratinocyte function [6, 29]. Interestingly, TSH did not affect basal epidermal keratinocyte proliferation *ex vivo*, pointing to the fact that the upregulation of K5 and K14 was not just due to enhanced keratinocyte proliferation. Therefore, these findings suggest that TSH effects on keratin expression are direct and independent of cellular proliferation changes. Just like with TRH, TSH was also found to enhance mitochondrial activity in the epidermis [27] and the HF epithelium [16], again raising the possibility of a coordinated, neurohormone-controlled increase in intraepithelial energy metabolism and keratin synthesis.

As alluded to above, keratin changes following TSH stimulation were also evident in human HFs *ex vivo*. Except for K5 in hair matrix keratinocytes, which was upregulated [18], all the other keratins examined were downregulated following TSH stimulation at the gene and protein levels. These included keratins expressed in the HF ORS, such as K6, K14, and K17, and hair keratins expressed in the hair cortex, such as K31, K32, and K85 [39]. While the exact mechanisms by which TSH changes keratin expression remains unknown, it is noteworthy that TSH also upregulated expression of MSX2 [39], a key transcription factor that controls keratin expression [40, 41]. It is also interesting to note that all these keratin changes were observed in the HF, although TSH itself does not affect hair growth, thus suggesting that these TSH-regulated changes in keratin expression do not translate into altered hair growth [18].

TRH has been found to enhance TSH expression in the human epidermis [13]. Since TSH can change keratin expression as we have just reviewed, it is possible that some of the effects of TRH on keratin expression are indirectly mediated by TSH. Indeed, some of the keratins that are modulated by TRH, such as K14, K17, and K85, are affected in a comparable manner by TSH [39].

3. The hypothalamus-pituitary-adrenal axis in the skin and its effects on keratin expression

Corticotropin-releasing hormone (CRH), adrenocorticotropic hormone (ACTH), and cortisol form the HPA axis, which has a major role in controlling stress response by producing steroid hormones and regulatory peptides [9]. This axis is also active in human skin and HFs, where, namely, keratinocytes, but also other cutaneous cell populations act as both targets and also as nonclassical producers of these HPA constituents [10–12, 42, 43].

There are plenty of studies that report on the effect of glucocorticosteroids on keratin expression in the skin, effects that accompany physiological processes, such as wound healing [6]. Nevertheless, little is known on the effects of the other components of the HPA axis on keratin expression, and the available information is limited to CRH, which reportedly upregulates K1 and downregulates K14 in HaCaT cells and in human adult epidermal keratinocytes, as part of the induction of the terminal differentiation program [44, 45]. Taking into consideration the fact that the HPA is fully functional in human skin [2, 10–12], it is likely that additional keratins are regulated by these neuromediators, yet have escaped notice so far. Therefore, further research is warranted to explore this neuroendocrine frontier of keratin biology, namely, in human skin.

4. Prolactin effects on keratin expression

Prolactin and its receptor have been found to be expressed at the gene and protein levels in the human skin [19, 35, 46, 47], where they control a large number of functions, such as hair growth [19] and keratin expression (see below). Given the major role of prolactin in the control of mammary development, growth, and milk production, it is not surprising that the first evidence for an effect of prolactin on keratin expression arose from mammary gland studies [48]. These studies have shown that if the prolactin receptor gene is knocked out, mice do not develop normal mammary buds, accompanied by decreased expression of selected keratins, such as K8, K17, K18, and K19 [48].

Since the mammary gland is basically a sweat gland-like derivative of the epidermis, and a prolactin-like protein has actually been found in human eccrine sweat glands [49], it was reasonable to hypothesize that prolactin may regulate keratin expression also in other skin appendages. Indeed, prolactin administration to organ-cultured human HFs resulted in upregulation of keratins expressed in the ORS, including K5 and K14, while the hair keratin K31 was downregulated *ex vivo* [50].

Perhaps the most interesting observation that emerged from this study was the stimulatory effect of prolactin on K15 and K19, that is, marker keratins for epithelial HF stem cells [51–53]. This stimulatory effect was reversed when a selective prolactin receptor antagonist was added to the culture medium. This effect was further confirmed when prolactin had a stimulatory effect on *KRT15* promoter activity *in situ* [50]. This finding strengthens the importance of prolactin as a stem cell promoting agent, as was also observed later in other classical prolactin target organs, such as the mammary gland [54]. Once again, this underscored the unique instructiveness of HFs as a discovery tool in skin research, namely, in cutaneous neuroendocrinology [2], from which novel, general neuroendocrine principles can be deduced.

Another important observation that emerged from these keratin studies was that the addition of a prolactin receptor antagonist alone also resulted in changes in keratin expression [50]. This shows that endogenous production of prolactin and/or prolactin receptor stimulation is an important element of normal skin physiology and homeostasis and is actually required to maintain the production of keratins in the HF. This is similar to the autocrine/paracrine effects attributed to prolactin also in the pituitary gland, where blocking of the prolactin receptor resulted in changes in cell turnover and prolactin receptor expression [55], and in extrapituitary locations such as the mammary gland, where changes in prolactin receptor patterning resulted in disruption of lobuloalveolar development [56].

It has been previously shown that there is an interplay between the different hormones and neurohormones in the skin and HFs, and that some of these connections are similar to those that exist in the pituitary. As an example, TRH can stimulate prolactin expression in the HF, while it can inhibit expression of the prolactin receptor [35]. Such an interplay is highly likely to also be at play in the regulation of keratin expression, and given that both neurohormones profoundly change the expression of selected keratins in human skin. Obviously, this adds another level of complexity to the challenge of segregating the direct effects of each of these neurohormones from indirect and cross-regulatory ones.

5. The effects of endocannabinoids on keratin expression

Accumulating data show that the endocannabinoid system (ECS) plays a major role in mammalian skin [57, 58]. Indeed, endocannabinoids are being produced by the epidermis and the skin appendages, including the HF, sweat glands, and sebaceous glands [58], and the cannabinoid receptors CB_1 and CB_2 are prominently expressed on different skin cell populations [58]. Many different skin functions of the skin are now appreciated to be regulated by the ECS. For example, in the epidermis, it controls keratinocyte proliferation and differentiation, thereby affecting the epidermal barrier, and regulates melanogenesis [59–61].

The ECS also affects the skin appendages profoundly. Signaling via CB_1 inhibits hair growth and induces catagen, the regression phase of the HF [22, 62]. In sweat glands, anandamide stimulated sweat secretion of epithelial cells and reduced their proliferation [63]. The ECS can also affect sebaceous gland function, and by acting via CB_2, endocannabinoids positively control sebaceous lipid synthesis [64]. Furthermore, cannabidiol, a CB1 antagonizing nonpsychotropic phytocannabinoid, reduced sebocyte proliferation and normalized excess sebum production that can be observed in acne lesions [65, 66].

Taking into consideration its importance in epidermal keratinocyte function, it was not surprising that ECS modulation also affects keratin expression. For example, cannabinoid receptor activation on human HaCaT cells by the prototypic endocannabinoid, anandamide, inhibited cell differentiation, accompanied by reduced transcription of the *KRT1* and *KRT10* genes [67]. When tested in human skin culture and again in HaCaT cells, anandamide also inhibited K6 and K16 expression, independent of its antiproliferative properties [68]. Conversely, administration of the CB_1 antagonist, arachidonyl-2′-chloroethylamide (ACEA), upregulated K10 in human epidermis while decreasing the expression of K1 *ex vivo* [69].

Given its antiproliferative and differentiation-promoting effects in human epidermis as well as its overall largely anti-inflammatory properties (e.g., by reducing mast cell degranulation and maturation *in loco* [70]), CB ligands are coming under scrutiny as potential new therapeutics in the therapy of psoriasis [71]. If this line of research continues to be productive, it will become clinically even more important to dissect the relative contribution of CB-mediated changes in epidermal keratin expression to any beneficial effects observed by therapeutic CB stimulation. The use of ECS antagonists to change keratin expression underscores that, like we have seen in the case of prolactin, blocking the autocrine/paracrine effects of intracutaneously generated neuroendocrine mediators induces functionally relevant changes in human skin, such as altered keratin expression patterns.

6. Opioids and keratin expression

Murine and human skin both express opioid receptors, including the μ-, κ-, and δ-opioid receptors. Stimulation of these receptors participates in the control of melanocyte [72] and keratinocyte functions, such as impeding DNA synthesis and cell differentiation [73, 74]. Therefore, their connection to skin disorders, such as psoriasis, basal cell carcinoma, and wound healing, is currently under scrutiny [73, 75, 76].

As one might expect by now, opioid receptor ligands also induce changes in keratin expression. For example, the key endogenous ligand for the μ-opiate receptor, beta-endorphin, enhances the intraepidermal expression of K16 at the wound margin [77]. In psoriasis, a hyperproliferative dermatosis, K16 expression is upregulated, and this is accompanied by downregulation of the μ-opiate receptor [75], and treatment of skin organ cultures with beta-endorphin resulted in elevated K16 production [75].

K10 is an additional keratin to be regulated by opioids, as mice knocked out for the δ-opioid receptor had enhanced K10 expression, together with a thinner epidermis [78], and the *Achillea millefolium* extract, a strong inducer of the μ-opioid receptor-1, led to increased differentiation of the cells in the epidermis with stronger K10 expression [79]. Yet, our current understanding of the role of opioid receptor-mediated signaling within the emerging neuroendocrine controls of keratin biology remains even more rudimentary than that of the neuromediators discussed further above.

7. Other neurohormones can alter keratin expression

Parathyroid hormone-related protein (PTHrP) is another important neuroendocrine mediator, which has importance in the normal formation of the mammary gland [80]. Keratin expression was tested in a K14 promoter-driven PTHrP mouse, and an overexpression of K17 in the nipple epidermis was evident in this mouse model [81]. Interestingly, PTHrP signaling affects BMP signaling and *Msx* gene activation, both of which are critical regulators of HF growth and function [80], just like PTHrP itself strongly modulates murine HF cycling [82, 83]. Yet, how PTHrP impacts on intrafollicular keratin remains to be evaluated.

Catecholamines can also change keratin expression, and when evaluated in limbal epithelial cells in culture, isoproterenol, a beta-adrenergic receptor agonist, led to pronounced changes in keratin expression [84]. When tested in HaCaT cells, the same compound stimulated differentiation, which was accompanied by increased K1 and K10 production [85].

In contrast, histamine led to decreased expression of differentiation markers in skin models and human keratinocyte cultures, among others, and also to decreased production of K1 and K10 [86]. The cholinergic system can also affect keratin expression. When tested in skin cultures *in vitro*, blocking of the cholinergic system resulted in decreased expression of differentiation markers, such as K2 and K10 [87]. Although these mediators clearly led to changes in keratin expression in these cases, it remains to be dissected whether these changes were due to a direct effect of the tested compound or reflected secondary events, resulting, for example, from changes in keratinocyte proliferation and differentiation.

8. Possible clinical implications of neuroendocrine-mediated changes in keratin expression

As reviewed in detail above, neuroendocrine mediators can change keratin expression in what appears to be a relatively selective manner. Let us now discuss, therefore, how this phenomenon might be translated into the treatment of several skin and hair conditions. This is of special clinical relevance since neuromediator analogs, in principle, may be formulated to be topically applicable, thus circumventing or reducing the risk of undesired systemic effects. Some of the possible clinical scenarios for which such analogs may conceivably be used are described briefly below.

8.1. Treatment of keratin-related skin and hair genetic disorders

The list of genetic disorders linked to mutations in keratin genes continues to expand, and more than half of the keratin genes have been linked to a genetic disorder [88–92]. These disorders include ichthyoses, blistering disorders such as epidermolysis bullosa, hair conditions such as wooly hair and sparse hair, and changes in the normal growth of nails. A novel promising approach for the treatment of keratin disorders is the utilization of small molecule drugs to upregulate expression of compensatory keratins or to downregulate the expression of the mutated keratins [89, 93]. Such an approach has already been successful in several autosomal dominant keratin disorders, such as epidermolysis bullosa simplex and pachyonychia congenita [94–96].

It has also been reported to be of potential benefit in epidermolytic ichthyosis, an uncommon genodermatosis caused by mutations in keratins 1 or 10, when Reichelt et al. have shown that increased stability of keratins 5 and 14 could lead to the formation of normal epidermis in K10-null mice [97]. Furthermore, treatment of immortalized cell lines from a *KRT10*-mutated epidermolytic ichthyosis patient with all-trans retinoic acid led to a 200-fold decrease in mRNA expression of K10, accompanied by decreased keratin aggregation [98].

As reviewed above, the CB_1 agonist ACEA increased K10 expression, while reducing K1 production in human epidermis in culture [69]. Such changes could potentially be harnessed in epidermolytic ichthyosis patients to decrease the expression of mutated K1 while upregulating the expression of K10 that can functionally compensate in part for the mutated keratin. Given their differential regulation of distinct human keratins in human skin *ex vivo*, defined neuromediators now need to be systematically explored for their capacity to execute such therapeutically desirable reverse regulation of clinically relevant keratins in selected genodermatoses, perhaps starting with primary keratinocyte cultures derived from affected patients.

8.2. Treatment of inflammatory skin conditions (e.g., psoriasis)

Several inflammatory skin disorders are characterized by overexpression of K6. These include, for example, lichen planus and discoid lupus erythematosus [99]. However, the most prominent example is psoriasis, a chronic inflammatory skin condition, which is characterized by increased expression of K6, K16, and K17 [3, 68, 100]. K17 is probably of special importance in psoriasis pathogenesis, since it has been suggested to act as an antigenic target for T lymphocytes in the affected epidermis [101]. Furthermore, mice overexpressing K17 developed an inflammatory reaction and epidermal hyperplasia [102]. Moreover, K6, K16, and K17 expression pattern can impact on the cytokine or chemokine secretion of keratinocytes [102–105] and thus the intraepidermal inflammatory signaling milieu.

Therefore, compounds that can decrease the expression of these keratins might be therapeutically beneficial in these dermatoses, namely, in psoriasis, especially if they can also exert anti-inflammatory effects [106, 107], such as in the case of cannabinoid receptor agonists, which independently decrease the expression of K6 and K16 [68], combined with anti-inflammatory, antiproliferative, and antiangiogenic properties [3, 57, 71, 108, 109].

8.3. Wound healing

In healthy nonglabrous epidermis, K6, K16, and K17 are largely absent and not constitutively expressed by keratinocytes. However, in hyperproliferative states and conditions of epidermal stress, such as during wound healing, these keratins are rapidly upregulated and strongly expressed, since they play a major role in epidermal repair, as they are required for normal migration of keratinocytes from the wound edges and to ensure optimal closure of the wound [29, 110, 111]. Opiate receptor agonists that can boost wound healing are also stimulators for K16 expression, suggesting again the hypothesis of a coordinated neuroendocrine control of both, expression of optimally suited keratins and wound healing as such [77, 112]. Conceivably, therefore, neuroendocrine mediators that upregulate K6, K16, and K17 expression (e.g., catecholamines and endocannabinoids) might become therapeutically useful as promoters of re-epithelialization during wound healing.

8.4. Therapeutic regulation of stem cell-associated keratins

Prolactin increases the expression of the prototypic epithelial stem/progenitor cell-associated keratins, K15 and K19, [48, 50], and a continuous endogenous production of prolactin may be required to maintain normal K15 and K19 expression by these stem cells [50]. This raises the question whether neurohormones such as prolactin or related receptor agonists can be therapeutically recruited to ameliorate or prevent stem cell-based hair diseases characterized by permanent loss of the HF stem cell pool, such as lichen planopilaris or chemotherapy-induced alopecia [51, 113–115], or epidermal atrophy associated with an exhaustion of epidermal stem cell pools, as it occurs, for example, in connection with steroid therapy [116].

8.5. Hair growth

Keratins play a critical role in normal hair growth and structure. This is nicely exemplified by genetic hair disorders caused by keratin mutations [91]. When keratins that are produced in the hair cortex are mutated, the hair shaft is fragile and easy to break, and when the mutations are in keratins expressed in the most proximal part of the hair cortex, this leads to a more severe phenotype of complete hair loss [117, 118]. Instead, when keratins expressed in the IRS are mutated, this leads to a defect in hair curvature, oftentimes evident as wooly hair [90, 92, 119–121]. It is therefore conceivable that neuroendocrine manipulation of hair keratin expression may result in modulation of hair growth and/or hair shaft phenotype. It is therefore not surprising that TRH and prolactin, which both significantly modulate hair growth [2, 3, 23, 50, 122–126], also profoundly modulate hair keratin expression [37, 50].

One additional important aspect when discussing hair keratins is the presence and importance of KAPs. These proteins surround the keratin intermediate filaments in the hair shaft, cross-linking them by disulfide bonds [127], and providing them with rigidity and strength [128]. The number of KAPs is much higher than keratins, and 89 functional KAP genes have been described in humans [128], therefore there is probably a high degree of overlap between these proteins. Nevertheless, changes in KAPs could probably also affect hair structure. On this background, it is interesting to note that preliminary studies using microarrays in cultured HFs have revealed that certain neurohormones, such as TSH and prolactin, appear to alter the transcription of several KAP genes, such as KAP 4-4 and/or KAP 7-1 [18, 50]. These pilot observations deserve systematic follow up and may provide additional targets for therapeutic neuroendocrine intervention.

9. Conclusions

Here, we have reviewed that several neurohormones and neuropeptides generated in human skin as a nonclassical production site profoundly impact on the control of keratin expression. Specifically, we have presented TRH, TSH, opioids, prolactin, and cannabinoid receptor ligands as prominent examples for and indicators of a likely much more widespread and complex, evolutionarily conserved neuroendocrine regulation of human keratin biology *in situ* than we have come to appreciate so far. We have argued that this regulation is critically involved in the maintenance of skin and HF homeostasis and may participate in skin pathology. Thus, it is timely that mainstream keratin and neuroendocrinology research, which traditionally interconnect only rarely, discover the cross-fertilization potential and clinical relevance of systematically exploring the neuroendocrine control of keratin expression and its functional consequences, namely, in human skin and HFs. Besides defining some of the many open questions in the field, we have provided specific examples for how neurohormones may be harnessed to treat selected genodermatoses and other skin disorders accompanied by abnormal keratin expression.

Many obstacles encumber the ongoing journey toward understanding mechanistically how exactly these neuromediators change keratin expression on the molecular level, and in uncovering which of these effects are directly or indirectly mediated (e.g., by affecting other cutaneous functions, which then impact on keratin expression). This situation has been further complicated by increasing insight into the strong interplay between and cross-regulation of different neurohormones within human skin. However, recent advances and refinements of serum-free human skin and HF organ cultures, which permits the silencing of selected neurohormone and receptor genes [70, 129], and the use of selective neurohormone receptor antagonists [50] surely facilitate progress in this exciting, translationally relevant line of investigation.

Conflict of interest

The authors declare they have no conflicts of interest. For the record, however, RP is founder of Monasterium Laboratory, Münster/Germany (www.monasteriumlab.com), a hair and skin research company, and consults for several companies with an interest in skin and hair research.

http://dx.doi.org/10.5772/intechopen.80406

Author details

Yuval Ramot[1]* and Ralf Paus[2,3,4]

*Address all correspondence to: yramot@gmail.com

1 Department of Dermatology, Hadassah—Hebrew University Medical Center, Jerusalem, Israel

2 Centre for Dermatology Research, The University of Manchester, Manchester, UK

3 NIHR Manchester Biomedical Research Centre and Manchester Academic Health Science Centre, Manchester, UK

4 Department of Dermatology and Cutaneous Surgery, University of Miami, Miller School of Medicine, Miami, USA

References

[1] Hsu CK, Lin HH, Harn HI, Hughes MW, Tang MJ, Yang CC. Mechanical forces in skin disorders. The Journal of Dermatological Science. 2018;**90**(3):232-240

[2] Paus R, Langan EA, Vidali S, Ramot Y, Andersen B. Neuroendocrinology of the hair follicle: Principles and clinical perspectives. Trends in Molecular Medicine. 2014;**20**(10): 559-570

[3] Ramot Y, Paus R. Harnessing neuroendocrine controls of keratin expression: A new therapeutic strategy for skin diseases? BioEssays. 2014;**36**(7):672-686

[4] Coulombe PA. The molecular revolution in cutaneous biology: Keratin genes and their associated disease: Diversity, opportunities, and challenges. The Journal of Investigative Dermatology. 2017;**137**(5):e67-e71

[5] Coulombe PA. Discovery of keratin function and role in genetic diseases: The year that 1991 was. Molecular Biology of the Cell. 2016;**27**(18):2807-2810

[6] Ramot Y, Paus R, Tiede S, Zlotogorski A. Endocrine controls of keratin expression. BioEssays. 2009;**31**(4):389-399

[7] Slominski AT, Hardeland R, Zmijewski MA, Slominski RM, Reiter RJ, Paus R. Melatonin: A cutaneous perspective on its production, metabolism, and functions. The Journal of Investigative Dermatology. 2018;**138**(3):490-499

[8] Slominski AT, Zmijewski MA, Plonka PM, Szaflarski JP, Paus R. How ultraviolet light touches the brain and endocrine system through skin, and why. Endocrinology. 2018;**159**(5):1992-2007

[9] Slominski AT, Zmijewski MA, Skobowiat C, Zbytek B, Slominski RM, Steketee JD. Sensing the environment: Regulation of local and global homeostasis by the skin's neuroendocrine system. Advances in Anatomy, Embryology and Cell Biology. 2012;**212**:v, vii, 1-115

[10] Slominski AT, Zmijewski MA, Zbytek B, Tobin DJ, Theoharides TC, Rivier J. Key role of CRF in the skin stress response system. Endocrine Reviews. 2013;**34**(6):827-884

[11] Ito N, Ito T, Kromminga A, Bettermann A, Takigawa M, Kees F, et al. Human hair follicles display a functional equivalent of the hypothalamic-pituitary-adrenal axis and synthesize cortisol. The FASEB Journal. 2005;**19**(10):1332-1334

[12] Slominski A, Wortsman J, Tuckey RC, Paus R. Differential expression of HPA axis homolog in the skin. Molecular and Cellular Endocrinology. 2007;**265-266**:143-149

[13] Bodo E, Kany B, Gaspar E, Knuver J, Kromminga A, Ramot Y, et al. Thyroid-stimulating hormone, a novel, locally produced modulator of human epidermal functions, is regulated by thyrotropin-releasing hormone and thyroid hormones. Endocrinology. 2010; **151**(4):1633-1642

[14] Slominski A, Wortsman J, Kohn L, Ain KB, Venkataraman GM, Pisarchik A, et al. Expression of hypothalamic-pituitary-thyroid axis related genes in the human skin. The Journal of Investigative Dermatology. 2002;**119**(6):1449-1455

[15] van Beek N, Bodo E, Kromminga A, Gaspar E, Meyer K, Zmijewski MA, et al. Thyroid hormones directly alter human hair follicle functions: Anagen prolongation and stimulation of both hair matrix keratinocyte proliferation and hair pigmentation. The Journal of Clinical Endocrinology and Metabolism. 2008;**93**(11):4381-4388

[16] Vidali S, Knuever J, Lerchner J, Giesen M, Biro T, Klinger M, et al. Hypothalamic-pituitary-thyroid axis hormones stimulate mitochondrial function and biogenesis in human hair follicles. The Journal of Investigative Dermatology. 2014;**134**(1):33-42

[17] Paus R, Theoharides TC, Arck PC. Neuroimmunoendocrine circuitry of the 'brain-skin connection'. Trends in Immunology. 2006;**27**(1):32-39

[18] Bodo E, Kromminga A, Biro T, Borbiro I, Gaspar E, Zmijewski MA, et al. Human female hair follicles are a direct, nonclassical target for thyroid-stimulating hormone. The Journal of Investigative Dermatology. 2009;**129**(5):1126-1139

[19] Foitzik K, Krause K, Conrad F, Nakamura M, Funk W, Paus R. Human scalp hair follicles are both a target and a source of prolactin, which serves as an autocrine and/or paracrine promoter of apoptosis-driven hair follicle regression. The American Journal of Pathology. 2006;**168**(3):748-756

[20] Foitzik K, Krause K, Nixon AJ, Ford CA, Ohnemus U, Pearson AJ, et al. Prolactin and its receptor are expressed in murine hair follicle epithelium, show hair cycle-dependent expression, and induce catagen. The American Journal of Pathology. 2003; **162**(5):1611-1621

[21] Gaspar E, Nguyen-Thi KT, Hardenbicker C, Tiede S, Plate C, Bodo E, et al. Thyrotropin-releasing hormone selectively stimulates human hair follicle pigmentation. The Journal of Investigative Dermatology. 2011;**131**(12):2368-2377

[22] Telek A, Biro T, Bodo E, Toth BI, Borbiro I, Kunos G, et al. Inhibition of human hair follicle growth by endo- and exocannabinoids. The FASEB Journal. 2007;**21**(13):3534-3541

[23] Gaspar E, Hardenbicker C, Bodo E, Wenzel B, Ramot Y, Funk W, et al. Thyrotropin releasing hormone (TRH): A new player in human hair-growth control. The FASEB Journal. 2010;**24**(2):393-403

[24] Paus R. A neuroendocrinological perspective on human hair follicle pigmentation. Pigment Cell and Melanoma Research. 2011;**24**(1):89-106

[25] Meier NT, Haslam IS, Pattwell DM, Zhang GY, Emelianov V, Paredes R, et al. Thyrotropin-releasing hormone (TRH) promotes wound re-epithelialisation in frog and human skin. PLoS One. 2013;**8**(9):e73596

[26] Knuever J, Poeggeler B, Gaspar E, Klinger M, Hellwig-Burgel T, Hardenbicker C, et al. Thyrotropin-releasing hormone controls mitochondrial biology in human epidermis. The Journal of Clinical Endocrinology and Metabolism. 2012;**97**(3):978-986

[27] Poeggeler B, Knuever J, Gaspar E, Biro T, Klinger M, Bodo E, et al. Thyrotropin powers human mitochondria. The FASEB Journal. 2010;**24**(5):1525-1531

[28] Paus R, Arck P. Neuroendocrine perspectives in alopecia areata: Does stress play a role? The Journal of Investigative Dermatology. 2009;**129**(6):1324-1326

[29] Moll R, Divo M, Langbein L. The human keratins: Biology and pathology. Histochemistry and Cell Biology. 2008;**129**(6):705-733

[30] Pan X, Hobbs RP, Coulombe PA. The expanding significance of keratin intermediate filaments in normal and diseased epithelia. Current Opinion in Cell Biology. 2013;**25**(1):47-56

[31] Eckhart L, Ehrlich F. Evolution of trichocyte keratins. Advances in Experimental Medicine and Biology. 2018;**1054**:33-45

[32] Kassem R, Liberty Z, Babaev M, Trau H, Cohen O. Harnessing the skin-thyroid connection for wound healing: A prospective controlled trial in guinea pigs. Clinical and Experimental Dermatology. 2012;**37**(8):850-856

[33] Safer JD, Crawford TM, Holick MF. Topical thyroid hormone accelerates wound healing in mice. Endocrinology. 2005;**146**(10):4425-4430

[34] Safer JD, Crawford TM, Holick MF. A role for thyroid hormone in wound healing through keratin gene expression. Endocrinology. 2004;**145**(5):2357-2361

[35] Langan EA, Ramot Y, Hanning A, Poeggeler B, Biro T, Gaspar E, et al. Thyrotropin-releasing hormone and oestrogen differentially regulate prolactin and prolactin receptor expression in female human skin and hair follicles in vitro. The British Journal of Dermatology. 2010;**162**(5):1127-1131

[36] Paus R. Exploring the "thyroid-skin connection": Concepts, questions, and clinical relevance. The Journal of Investigative Dermatology. 2010;**130**(1):7-10

[37] Ramot Y, Zhang G, Biro T, Langbein L, Paus R. Is thyrotropin-releasing hormone a novel neuroendocrine modulator of keratin expression in human skin? The British Journal of Dermatology. 2013;**169**(1):146-151

[38] Cianfarani F, Baldini E, Cavalli A, Marchioni E, Lembo L, Teson M, et al. TSH receptor and thyroid-specific gene expression in human skin. The Journal of Investigative Dermatology. 2010;**130**(1):93-101

[39] Ramot Y, Zhang G, Biro T, Lisztes E, Funk W, Ingber A, et al. TSH is a novel neuroendocrine regulator of selected keratins in the human hair follicle. Journal of Dermatological Science. 2011;**64**(1):67-70

[40] Cai J, Lee J, Kopan R, Ma L. Genetic interplays between Msx2 and Foxn1 are required for Notch1 expression and hair shaft differentiation. Developmental Biology. 2009;**326**(2): 420-430

[41] Cai J, Ma L. Msx2 and Foxn1 regulate nail homeostasis. Genesis. 2011;**49**(6):449-459

[42] Skobowiat C, Postlethwaite AE, Slominski AT. Skin exposure to ultraviolet B rapidly activates systemic neuroendocrine and immunosuppressive responses. Photochemistry and Photobiology. 2017;**93**(4):1008-1015

[43] Slominski AT, Zmijewski MA, Plonka PM, Szaflarski JP, Paus R. How UV light touches the brain and endocrine system through skin, and why. Endocrinology. 2018;**159**(5): 1992-2007

[44] Zbytek B, Pikula M, Slominski RM, Mysliwski A, Wei E, Wortsman J, et al. Corticotropin-releasing hormone triggers differentiation in HaCaT keratinocytes. The British Journal of Dermatology. 2005;**152**(3):474-480

[45] Zbytek B, Slominski AT. Corticotropin-releasing hormone induces keratinocyte differentiation in the adult human epidermis. Journal of Cellular Physiology. 2005;**203**(1):118-126

[46] Langan EA, Vidali S, Pigat N, Funk W, Lisztes E, Biro T, et al. Tumour necrosis factor alpha, interferon gamma and substance P are novel modulators of extrapituitary prolactin expression in human skin. PLoS One. 2013;**8**(4):e60819

[47] Richards RG, Hartman SM. Human dermal fibroblast cells express prolactin in vitro. The Journal of Investigative Dermatology. 1996;**106**(6):1250-1255

[48] Ormandy CJ, Naylor M, Harris J, Robertson F, Horseman ND, Lindeman GJ, et al. Investigation of the transcriptional changes underlying functional defects in the mammary glands of prolactin receptor knockout mice. Recent Progress in Hormone Research. 2003;**58**:297-323

[49] Walker AM, Robertson MT, Jones CJ. Distribution of a prolactinlike material in human eccrine sweat glands. The Journal of Investigative Dermatology. 1989;**93**(1):50-53

[50] Ramot Y, Biro T, Tiede S, Toth BI, Langan EA, Sugawara K, et al. Prolactin—A novel neuroendocrine regulator of human keratin expression in situ. The FASEB Journal. 2010; **24**(6):1768-1779

[51] Purba TS, Brunken L, Hawkshaw NJ, Peake M, Hardman J, Paus R. A primer for studying cell cycle dynamics of the human hair follicle. Experimental Dermatology. 2016;**25**(9):663-668

[52] Purba TS, Haslam IS, Poblet E, Jimenez F, Gandarillas A, Izeta A, et al. Human epithelial hair follicle stem cells and their progeny: Current state of knowledge, the widening gap in translational research and future challenges. BioEssays. 2014;**36**(5):513-525

[53] Purba TS, Peake M, Farjo B, Farjo N, Bhogal RK, Jenkins G, et al. Divergent proliferation patterns of distinct human hair follicle epithelial progenitor niches in situ and their differential responsiveness to prostaglandin D2. Scientific Reports. 2017;**7**(1):15197

[54] O'Leary KA, Shea MP, Salituro S, Blohm CE, Schuler LA. Prolactin alters the mammary epithelial hierarchy, increasing progenitors and facilitating ovarian steroid action. Stem Cell Reports. 2017;**9**(4):1167-1179

[55] Ferraris J, Boutillon F, Bernadet M, Seilicovich A, Goffin V, Pisera D. Prolactin receptor antagonism in mouse anterior pituitary: Effects on cell turnover and prolactin receptor expression. American Journal of Physiology. Endocrinology and Metabolism. 2012;**302**(3):E356-E364

[56] Grimm SL, Seagroves TN, Kabotyanski EB, Hovey RC, Vonderhaar BK, Lydon JP, et al. Disruption of steroid and prolactin receptor patterning in the mammary gland correlates with a block in lobuloalveolar development. Molecular Endocrinology. 2002; **16**(12):2675-2691

[57] Biro T, Toth BI, Hasko G, Paus R, Pacher P. The endocannabinoid system of the skin in health and disease: Novel perspectives and therapeutic opportunities. Trends in Pharmacological Sciences. 2009;**30**(8):411-420

[58] Maccarrone M, Bab I, Biro T, Cabral GA, Dey SK, Di Marzo V, et al. Endocannabinoid signaling at the periphery: 50 years after THC. Trends in Pharmacological Sciences. 2015; **36**(5):277-296

[59] Pucci M, Pasquariello N, Battista N, Di Tommaso M, Rapino C, Fezza F, et al. Endocannabinoids stimulate human melanogenesis via type-1 cannabinoid receptor. The Journal of Biological Chemistry. 2012;**287**(19):15466-15478

[60] Pucci M, Pirazzi V, Pasquariello N, Maccarrone M. Endocannabinoid signaling and epidermal differentiation. European Journal of Dermatology. 2011;**21**(Suppl 2):29-34

[61] Roelandt T, Heughebaert C, Bredif S, Giddelo C, Baudouin C, Msika P, et al. Cannabinoid receptors 1 and 2 oppositely regulate epidermal permeability barrier status and differentiation. Experimental Dermatology. 2012;**21**(9):688-693

[62] Srivastava BK, Soni R, Patel JZ, Joharapurkar A, Sadhwani N, Kshirsagar S, et al. Hair growth stimulator property of thienyl substituted pyrazole carboxamide derivatives as a CB1 receptor antagonist with in vivo antiobesity effect. Bioorganic and Medicinal Chemistry Letters. 2009;**19**(9):2546-2550

[63] Czifra G, Szollosi AG, Toth BI, Demaude J, Bouez C, Breton L, et al. Endocannabinoids regulate growth and survival of human eccrine sweat gland-derived epithelial cells. The Journal of Investigative Dermatology. 2012;**132**(8):1967-1976

[64] Dobrosi N, Toth BI, Nagy G, Dozsa A, Geczy T, Nagy L, et al. Endocannabinoids enhance lipid synthesis and apoptosis of human sebocytes via cannabinoid receptor-2-mediated signaling. The FASEB Journal. 2008;**22**(10):3685-3695

[65] Olah A, Markovics A, Szabo-Papp J, Szabo PT, Stott C, Zouboulis CC, et al. Differential effectiveness of selected non-psychotropic phytocannabinoids on human sebocyte functions implicates their introduction in dry/seborrhoeic skin and acne treatment. Experimental Dermatology. 2016;**25**(9):701-707

[66] Olah A, Toth BI, Borbiro I, Sugawara K, Szollosi AG, Czifra G, et al. Cannabidiol exerts sebostatic and antiinflammatory effects on human sebocytes. The Journal of Clinical Investigation. 2014;**124**(9):3713-3724

[67] Paradisi A, Pasquariello N, Barcaroli D, Maccarrone M. Anandamide regulates keratinocyte differentiation by inducing DNA methylation in a CB1 receptor-dependent manner. The Journal of Biological Chemistry. 2008;**283**(10):6005-6012

[68] Ramot Y, Sugawara K, Zakany N, Toth BI, Biro T, Paus R. A novel control of human keratin expression: Cannabinoid receptor 1-mediated signaling down-regulates the expression of keratins K6 and K16 in human keratinocytes in vitro and in situ. PeerJ. 2013;**1**:e40

[69] Ramot Y, Olah A, Paus R. Cover image: Neuroendocrine treatment of inherited keratin disorders by cannabinoids? The British Journal of Dermatology. 2018;**178**(6):1469

[70] Sugawara K, Biro T, Tsuruta D, Toth BI, Kromminga A, Zakany N, et al. Endocannabinoids limit excessive mast cell maturation and activation in human skin. The Journal of Allergy and Clinical Immunology. 2012;**129**(3):726-738 e8

[71] Derakhshan N, Kazemi M. Cannabis for refractory psoriasis-high hopes for a novel treatment and a literature review. Current Clinical Pharmacology. 2016;**11**(2):146-147

[72] Kauser S, Thody AJ, Schallreuter KU, Gummer CL, Tobin DJ. Beta-endorphin as a regulator of human hair follicle melanocyte biology. The Journal of Investigative Dermatology. 2004;**123**(1):184-195

[73] Bigliardi-Qi M, Bigliardi PL, Buchner S, Rufli T. Characterization of mu-opiate receptor in human epidermis and keratinocytes. Annals of the New York Academy of Sciences. 1999;**885**:368-371

[74] Tominaga M, Ogawa H, Takamori K. Possible roles of epidermal opioid systems in pruritus of atopic dermatitis. The Journal of Investigative Dermatology. 2007;**127**(9):2228-2235

[75] Bigliardi-Qi M, Bigliardi PL, Eberle AN, Buchner S, Rufli T. B-endorphin stimulates cytokeratin 16 expression and downregulates mu-opiate receptor expression in human epidermis. The Journal of Investigative Dermatology. 2000;**114**(3):527-532

[76] Kohl A, Werner A, Buntrock P, Diezel W, Adrian K, Titov MI. The effect of the peptide dalargin on wound healing. Dermatologische Monatschrift. 1989;**175**(9):561-572

[77] Bigliardi PL, Sumanovski LT, Buchner S, Rufli T, Bigliardi-Qi M. Different expression of mu-opiate receptor in chronic and acute wounds and the effect of beta-endorphin on transforming growth factor beta type II receptor and cytokeratin 16 expression. The Journal of Investigative Dermatology. 2003;**120**(1):145-152

[78] Bigliardi-Qi M, Gaveriaux-Ruff C, Zhou H, Hell C, Bady P, Rufli T, et al. Deletion of delta-opioid receptor in mice alters skin differentiation and delays wound healing. Differentiation. 2006;**74**(4):174-185

[79] Pain S, Altobelli C, Boher A, Cittadini L, Favre-Mercuret M, Gaillard C, et al. Surface rejuvenating effect of *Achillea millefolium* extract. International Journal of Cosmetic Science. 2011;**33**(6):535-542

[80] Hens JR, Dann P, Zhang JP, Harris S, Robinson GW, Wysolmerski J. BMP4 and PTHrP interact to stimulate ductal outgrowth during embryonic mammary development and to inhibit hair follicle induction. Development. 2007;**134**(6):1221-1230

[81] Eastwood J, Offutt C, Menon K, Keel M, Hrncirova P, Novotny MV, et al. Identification of markers for nipple epidermis: Changes in expression during pregnancy and lactation. Differentiation. 2007;**75**(1):75-83

[82] Peters EM, Foitzik K, Paus R, Ray S, Holick MF. A new strategy for modulating chemotherapy-induced alopecia, using PTH/PTHrP receptor agonist and antagonist. The Journal of Investigative Dermatology. 2001;**117**(2):173-178

[83] Schilli MB, Ray S, Paus R, Obi-Tabot E, Holick MF. Control of hair growth with parathyroid hormone [7-34]. The Journal of Investigative Dermatology. 1997;**108**(6):928-932

[84] Ghoubay-Benallaoua D, Pecha F, Goldschmidt P, Fialaire-Legendre A, Chaumeil C, Laroche L, et al. Effects of isoproterenol and cholera toxin on human limbal epithelial cell cultures. Current Eye Research. 2012;**37**(7):644-653

[85] Mammone T, Marenus K, Maes D, Lockshin RA. The induction of terminal differentiation markers by the cAMP pathway in human HaCaT keratinocytes. Skin Pharmacology and Applied Skin Physiology. 1998;**11**(3):152-160

[86] Gschwandtner M, Mildner M, Mlitz V, Gruber F, Eckhart L, Werfel T, et al. Histamine suppresses epidermal keratinocyte differentiation and impairs skin barrier function in a human skin model. Allergy. 2013;**68**(1):37-47

[87] Kurzen H, Henrich C, Booken D, Poenitz N, Gratchev A, Klemke CD, et al. Functional characterization of the epidermal cholinergic system in vitro. The Journal of Investigative Dermatology. 2006;**126**(11):2458-2472

[88] Chamcheu JC, Siddiqui IA, Syed DN, Adhami VM, Liovic M, Mukhtar H. Keratin gene mutations in disorders of human skin and its appendages. Archives of Biochemistry and Biophysics. 2011;**508**(2):123-137

[89] Chamcheu JC, Wood GS, Siddiqui IA, Syed DN, Adhami VM, Teng JM, et al. Progress towards genetic and pharmacological therapies for keratin genodermatoses: Current perspective and future promise. Experimental Dermatology. 2012;**21**(7):481-489

[90] Ramot Y, Zlotogorski A. Molecular genetics of alopecias. Current Problems in Dermatology. 2015;**47**:87-96

[91] Ramot Y, Zlotogorski A. Keratins: The hair shaft's backbone revealed. Experimental Dermatology. 2015;**24**(6):416-417

[92] Ramot Y, Zlotogorski A. The twisting tale of woolly hair: A trait with many causes. Journal of Medical Genetics. 2015;**52**(4):217-223

[93] Wally V, Lettner T, Peking P, Peckl-Schmid D, Murauer EM, Hainzl S, et al. The pathogenetic role of IL-1beta in severe epidermolysis bullosa simplex. The Journal of Investigative Dermatology. 2013;**133**(7):1901-1903

[94] Hickerson RP, Leake D, Pho LN, Leachman SA, Kaspar RL. Rapamycin selectively inhibits expression of an inducible keratin (K6a) in human keratinocytes and improves symptoms in pachyonychia congenita patients. Journal of Dermatological Science. 2009;**56**(2):82-88

[95] Kerns ML, DePianto D, Dinkova-Kostova AT, Talalay P, Coulombe PA. Reprogramming of keratin biosynthesis by sulforaphane restores skin integrity in epidermolysis bullosa simplex. Proceedings of the National Academy of Sciences of the United States of America. 2007;**104**(36):14460-14465

[96] Zhao Y, Gartner U, Smith FJ, McLean WH. Statins downregulate K6a promoter activity: A possible therapeutic avenue for pachyonychia congenita. The Journal of Investigative Dermatology. 2011;**131**(5):1045-1052

[97] Reichelt J, Bussow H, Grund C, Magin TM. Formation of a normal epidermis supported by increased stability of keratins 5 and 14 in keratin 10 null mice. Molecular Biology of the Cell. 2001;**12**(6):1557-1568

[98] Li H, Torma H. Retinoids reduce formation of keratin aggregates in heat-stressed immortalized keratinocytes from an epidermolytic ichthyosis patient with a KRT10 mutation. Acta Dermato-Venereologica. 2013;**93**(1):44-49

[99] Ichikawa E, Watanabe S, Takahashi H. Keratin and involucrin expression in discoid lupus erythematosus and lichen planus. Archives of Dermatological Research. 1997; **289**(9):519-526

[100] Ramot Y. Psoriasis and osteoporosis: The debate continues. The British Journal of Dermatology. 2017;**176**(5):1117-1118

[101] Gudmundsdottir AS, Sigmundsdottir H, Sigurgeirsson B, Good MF, Valdimarsson H, Jonsdottir I. Is an epitope on keratin 17 a major target for autoreactive T lymphocytes in psoriasis? Clinical and Experimental Immunology. 1999;**117**(3):580-586

[102] Depianto D, Kerns ML, Dlugosz AA, Coulombe PA. Keratin 17 promotes epithelial proliferation and tumor growth by polarizing the immune response in skin. Nature Genetics. 2010;**42**(10):910-914

[103] Chung BM, Arutyunov A, Ilagan E, Yao N, Wills-Karp M, Coulombe PA. Regulation of C-X-C chemokine gene expression by keratin 17 and hnRNP K in skin tumor keratinocytes. The Journal of Cell Biology. 2015;**208**(5):613-627

[104] Hobbs RP, Batazzi AS, Han MC, Coulombe PA. Loss of keratin 17 induces tissue-specific cytokine polarization and cellular differentiation in HPV16-driven cervical tumorigenesis in vivo. Oncogene. 2016;**35**(43):5653-5662

[105] Hobbs RP, Lessard JC, Coulombe PA. Keratin intermediate filament proteins - novel regulators of inflammation and immunity in skin. Journal of Cell Science. 2012;**125**(Pt 22): 5257-5258

[106] Chang T, Sun L, Wang Y, Wang D, Li W, Li C, et al. Inhibition of keratin 17 expression with antisense and RNAi strategies: Exploring novel therapy for psoriasis. Experimental Dermatology. 2011;**20**(7):555-560

[107] Fu M, Wang G. Keratin 17 as a therapeutic target for the treatment of psoriasis. Journal of Dermatological Science. 2012;**67**(3):161-165

[108] Norooznezhad AH, Norooznezhad F. Cannabinoids: Possible agents for treatment of psoriasis via suppression of angiogenesis and inflammation. Medical Hypotheses. 2017;**99**:15-18

[109] Toth BI, Dobrosi N, Dajnoki A, Czifra G, Olah A, Szollosi AG, et al. Endocannabinoids modulate human epidermal keratinocyte proliferation and survival via the sequential engagement of cannabinoid receptor-1 and transient receptor potential vanilloid-1. The Journal of Investigative Dermatology. 2011;**131**(5):1095-1104

[110] Pastar I, Stojadinovic O, Yin NC, Ramirez H, Nusbaum AG, Sawaya A, et al. Epithelialization in wound healing: A comprehensive review. Advances in Wound Care (New Rochelle). 2014;**3**(7):445-464

[111] Rotty JD, Coulombe PA. A wound-induced keratin inhibits Src activity during keratinocyte migration and tissue repair. The Journal of Cell Biology. 2012;**197**(3):381-389

[112] Stein C, Kuchler S. Targeting inflammation and wound healing by opioids. Trends in Pharmacological Sciences. 2013;**34**(6):303-312

[113] Harries MJ, Jimenez F, Izeta A, Hardman J, Panicker SP, Poblet E, et al. Lichen planopilaris and frontal fibrosing alopecia as model epithelial stem cell diseases. Trends in Molecular Medicine. 2018;**24**(5):435-448

[114] Ramot Y, Mastrofrancesco A, Camera E, Desreumaux P, Paus R, Picardo M. The role of PPARgamma-mediated signalling in skin biology and pathology: New targets and opportunities for clinical dermatology. Experimental Dermatology. 2015;**24**(4):245-251

[115] Paus R, Haslam IS, Sharov AA, Botchkarev VA. Pathobiology of chemotherapy-induced hair loss. The Lancet Oncology. 2013;**14**(2):e50-e59

[116] Chebotaev DV, Yemelyanov AY, Lavker RM, Budunova IV. Epithelial cells in the hair follicle bulge do not contribute to epidermal regeneration after glucocorticoid-induced cutaneous atrophy. The Journal of Investigative Dermatology. 2007;**127**(12):2749-2758

[117] van Steensel M, Vreeburg M, Urbina MT, Lopez P, Morice-Picard F, van Geel M. Novel KRT83 and KRT86 mutations associated with monilethrix. Experimental Dermatology. 2015;**24**(3):222-224

[118] Naeem M, Wajid M, Lee K, Leal SM, Ahmad W. A mutation in the hair matrix and cuticle keratin KRTHB5 gene causes ectodermal dysplasia of hair and nail type. Journal of Medical Genetics. 2006;**43**(3):274-279

[119] Horev L, Babay S, Ramot Y, Saad-Edin B, Moorad S, Ingber A, et al. Mutations in two genes on chromosome 13 resulting in a complex hair and skin phenotype due to two rare genodermatoses: KLICK and autosomal recessive woolly hair/hypotrichosis simplex. The British Journal of Dermatology. 2011;**164**(5):1113-1116

[120] Molho-Pessach V, Sheffer S, Siam R, Tams S, Siam I, Awwad R, et al. Two novel homozygous desmoplakin mutations in carvajal syndrome. Pediatric Dermatology. 2015; **32**(5):641-646

[121] Ramot Y, Molho-Pessach V, Meir T, Alper-Pinus R, Siam I, Tams S, et al. Mutation in KANK2, encoding a sequestering protein for steroid receptor coactivators, causes keratoderma and woolly hair. Journal of Medical Genetics. 2014;**51**(6):388-394

[122] Foitzik K, Langan EA, Paus R. Prolactin and the skin: A dermatological perspective on an ancient pleiotropic peptide hormone. The Journal of Investigative Dermatology. 2009;**129**(5):1071-1087

[123] Langan EA, Foitzik-Lau K, Goffin V, Ramot Y, Paus R. Prolactin: An emerging force along the cutaneous-endocrine axis. Trends in Endocrinology and Metabolism. 2010; **21**(9):569-577

[124] Langan EA, Griffiths CE, Paus R. Utilizing the hair follicle to dissect the regulation and autocrine/paracrine activities of prolactin in humans. American Journal of Physiology. Endocrinology and Metabolism. 2012;**302**(10):E1311-E1312

[125] Langan EA, Ramot Y, Goffin V, Griffiths CE, Foitzik K, Paus R. Mind the (gender) gap: Does prolactin exert gender and/or site-specific effects on the human hair follicle? The Journal of Investigative Dermatology. 2010;**130**(3):886-891

[126] Paus R. Exploring the "brain-skin connection": Leads and lessons from the hair follicle. Current Research in Translational Medicine. 2016;**64**(4):207-214

[127] Marshall RC, Orwin DF, Gillespie JM. Structure and biochemistry of mammalian hard keratin. Electron Microscopy Reviews. 1991;**4**(1):47-83

[128] Gong H, Zhou H, McKenzie GW, Yu Z, Clerens S, Dyer JM, et al. An updated nomenclature for keratin-associated proteins (KAPs). International Journal of Biological Sciences. 2012;**8**(2):258-264

[129] Samuelov L, Sprecher E, Tsuruta D, Biro T, Kloepper JE, Paus R. P-cadherin regulates human hair growth and cycling via canonical Wnt signaling and transforming growth factor-beta2. The Journal of Investigative Dermatology. 2012;**132**(10):2332-2341

Chapter 5

Keratins in Skin Epidermal Development and Diseases

Ling-juan Zhang

Additional information is available at the end of the chapter

http://dx.doi.org/10.5772/intechopen.79050

Abstract

Epidermal keratinocyte (KC), the major cell type in the skin epidermis, plays critical roles in forming a permeability barrier to separate internal organs from external stimuli. Keratins, constituting about 30–80% of the total protein in KCs, form the major intermediate filament cytoskeleton of KC. Keratins consist of 54 unique genes in humans and they are expressed in cell-, differentiation- and development-dependent manner. While keratin pairs K5-K14 and K1-K10 are normally associated with KCs at different cell differentiation stages, other keratin pairs such as K6-K16/K17 and K8–K18 and are usually not expressed in normal skin interfollicular epidermis, but are elevated during wounding, inflammatory skin diseases such as psoriasis or malignant conversion of KC. The expression and function of keratins are tightly regulated at both transcriptional and post-transcriptional levels. Inherited or spontaneous mutations in keratins or abnormal keratin regulations or modifications can cause KC and cutaneous tissue fragility, skin hypertrophic and inflammatory conditions or malignant transformation of KC, therefore accounting for a large number of disorders in human skin. Here we review the recent literature on how keratins are normally expressed during skin development and how mutations or misregulations of these keratins are involved in the pathogenesis of skin diseases.

Keywords: keratins, keratinocytes, skin barrier, differentiation, skin blistering diseases, wounding, psoriasis, skin tumor

1. Introduction: keratins—principal structural proteins of the skin epidermis

Locating at the outermost layer of the skin, the epidermis plays a critical role protecting our body against environmental pathogens and insults by forming a physical and immunological barrier. This protective role of skin epidermis is manifested by extensive cytoskeletal architecture, and keratins represent its principle structural protein, contributing to 30–80% of the total protein and forming the major intermediate filament cytoskeleton of the epidermis. Since keratin family proteins were initially characterized based on their mobility on 2D SDS-PAGE back in 1980s, more than 50 mammalian keratins have been identified and characterized. Keratins can be sub-classified into two distinct classes: Type I keratins, including K9–K40, are relative acidic (pKi = 4.5–5.5) and small (40–56.5 kDa) whereas type II keratins, including K1–K8 and K71–86), are more basic (pKi = 5.5–7.5) and larger (53–67 kDa) [1–3]. The active keratin genes are clustered into two dense region of the chromosome: all type II keratins plus one type I keratin (K18) are located on chromosome 12q, and the remaining type I keratins are all on chromosome 17q. Despite the fact that type I and type II keratins are located at distinct region of the chromosome, they show beautifully specific patterns of gene expression within adjacent epidermal cell layers and a specific pair of keratins are usually co-expressed as a heterodimer between one acidic (type I) and one basic (type II) keratin [4]. These keratin heterodimers self-assemble into antiparallel, staggered tetramers, yielding intermediate filament through lateral and longitudinal interactions [4].

The primary function of the keratin intermediate filament cytoskeleton is to provide cells with structural resilience against mechanical trauma, and this is especially important for epidermal cells because as the outermost barrier tissue the epidermis has to have the ability to resist some of the most severe physical stress levels experienced by any human tissue. The basic structural organization of keratin intermediate filaments contains four helical or coiled-coil segments flanked by N- and C-terminal glycine rich sequence. And most disease-causing mutations occur in the well-conserved coiled-coil domains of keratins, leading to disruption of secondary structure formation between the heterodimer and the subsequent aggregation of the keratin filaments. In addition to structural and mechanical support, cell-specific keratin expression also modulates growth, adhesion, migration and invasion of epithelial cells. Thus, dysfunction or mutations of keratin proteins are associated with a remarkable variety of skin disorders, such as skin blistering, inflammatory disorders and skin tumors. So far more than 100 different disorders (termed as keratinopathies) have been linked to inherited keratin changes (www.interfil.org).

In this chapter we will first describe how keratins are normally expressed and regulated during epidermal development and in adult homeostasis, then we will describe how mutation or abnormal expression or modification of keratin proteins are associated with various skin diseases, and their function and regulatory mechanisms during disease pathogenesis. Keratin mutations and related diseases in skin appendages, including hair and nail, will not be reviewed here.

2. Keratin expression during epidermal development

2.1. Keratin expression kinetics during epidermal development

The barrier function of epidermis is mainly provided by keratinocytes (KC), the predominant cell type in the epidermis, and it is maintained by a tightly controlled balance between proliferation and differentiation of KC [5, 6]. As shown in **Figure 1**, the murine epidermal development begins at embryonic day (E) 8.5 from a single layer of progenitor cells that express keratin pair K8–K18, and around E 9.5 these progenitor cells are specified to an epidermal cell fate and switch to express a different pair of keratins, K5–K14 [7–10]. Between E 10.5–E15.5, the committed KC begin a process of upward stratification and differentiation leading to the generation of suprabasal cells, in which cell cycle is arrested and early differentiation program is initiated and K5/K14 are substituted by K1 and K10. Studies have suggested that the ectopic expression of K10 inhibits cell cycle progression and proliferation and thus promotes terminal differentiation of keratinocytes [11, 12]. At E 16.5, the suprabasal KC commit to terminal differentiation and continue to migrate upward forming the granular layer, and these granular KCs express late differentiation markers, such as Filaggrin (FLG), Loricrin (LOR) and Involucrin (INV). By E 18.5, epidermis becomes fully mature, and terminally differentiated KCs become enucleated corneocytes forming the outer most cornified layer with complete barrier function.

Note that K15 is an additional type I keratin protein co-expressed with K5/K14 in the basal keratinocytes of stratified epithelia and in the bulge in hair follicles [13], and K2 is an additional type II keratin protein co-expressed with K1/K10 in the differentiated keratinocytes [14]. Although K15 or K2 are generally considered as minor keratins, but the ratio of K15 to K14 or K2 to K1 can vary dramatically during development or upon disease condition at different skin sites [15]. In adult skin, K15 expression is restricted to hair follicle stem cells [16, 17]. This cell-stage specific expression pattern of keratins is precisely maintained through postnatal development and adulthood under homeostatic condition in both human and mouse skin epidermis. Therefore K5 and/or K14 are highly specific markers for basal proliferative KC, K15 is generally used as a marker for epithelial stem cells in the hair follicle bulge, and K1 and/or K10 have been robustly used to mark KC in the early differentiation stage (**Figure 2**).

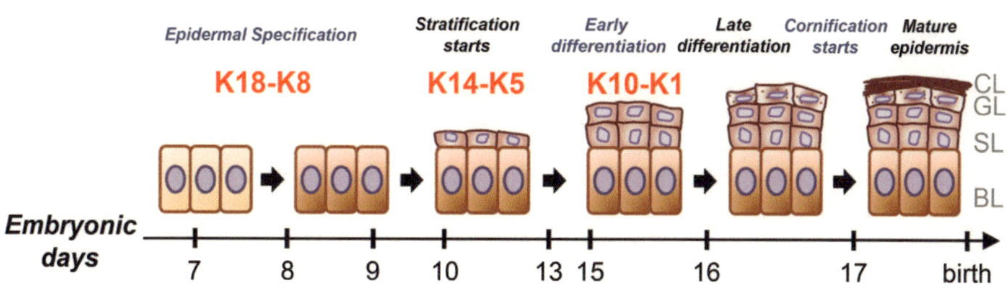

Figure 1. Overview of keratin expressions during different stage of mouse epidermal development. Murine epidermis develops from a single layer ectodermal keratinocyte (KC) progenitor cells which commit to an epidermal fat around embryonic day 8–9 (E 8–9). These KC progenitors express simple epithelia cell marker keratin pair of K18-K8. Around E 10.5 and the expression of K18/K8 starts to be substituted by K14/K5 in the committed KC, and stratification process starts leading to the formation of suprabasal cells. Around E15.5, early differentiation starts in the spinous/suprabasal layer (SL) and K10/K1 are expressed in the differentiated suprabasal cells while the expression of K14/K5 is restricted to basal layer (BL). At E16.5, suprabasal cells committed to terminal differentiation move upwards to form granular layer (GL) and K10/K1 expression are replaced by late differentiation markers. Finally cornification starts around E 16.5–17.5 and by E 18.5 epidermis becomes fully mature with intact cornified layer and complete barrier function.

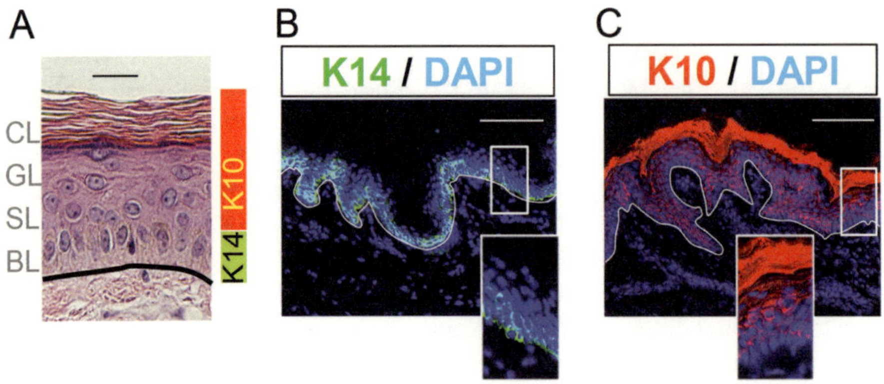

Figure 2. Expression of K14 and K10 in adult human skin epidermis. (A). H&E staining of adult human skin epidermis showing distinct epidermal layers including basal layer (BL), spinous (suprabasal) layer (SL), granular layer (GL) and cornified layer (CL). The dashed line marks the epidermal-dermal junction. Boxes on the right panel indicate the location of K14 and K10 expression in distinct epidermal layers. While K14 is restricted to cells in the basal layer, K10 is induced in early differentiated cells at the spinous layer and is maintained through all differentiated layers. Scale bar = 20 μm. (B-C). Adult human skin epidermis was immunostained by antibody specific for K14 (green in B) or K10 (red in C). Zoom-in pictures highlighted in white box are shown to illustrate distinct localization of K14 and K10 in skin epidermis. Scale bars = 100 μm.

2.2. Regulation of keratin expression during development

Keratin synthesis is regulated at the level of transcription by a characteristic constellation of transcription factors. Regulatory mechanisms for the expression of K5 and K14, the key keratins forming the cytoskeletal IF network in mitotically active basal cells, have been extensive studied. Studies have shown that the promoter activities of K5 and K14 genes are collaboratively regulated by several transcription factors, including AP1, AP2, NFκB, Skn-1a, Tst-1, RAR (nuclear receptor for retinoic acid), T3R (receptor for thyroid hormone), GR (glucocorticoid receptor) and coactivator CBP/p300 in response to many extracellular signals, such as growth factor, vitamins (retinoic acid/VitA), thyroid hormone, or glucocorticoids [18–22] .

The expression of K1 and K10 is upregulated during early differentiation process. Forced expression of transcription factor C/EBPβ in keratinocytes arrested growth and induced the expression of K1 and K10 but had a minimal effect on the expression of late differentiation markers [23]. In consistent, Cebpb-deficient mice had reduced levels of K1 and K10 but not of Lor or Inv, suggesting that C/EBPβ modulates K1 and K10 expression during early events of keratinocyte differentiation. Other transcription factors, such as C/EBPα and AP2 are also required for K10 expression during differentiation [24]. Therefore, C/EBPβ, C/EBPα and AP2 are considered the differentiation-associated transcription factors controlling early differentiation process of keratinocytes. We have shown previously that in cultured primary mouse basal keratinocytes, while elevating extracellular calcium robustly triggers stratification and the expression of late differentiation markers, K10 expression is only moderately increased by high calcium [6, 25]. In contrast, growth factor depletion/starvation strongly induced K10 expression [6, 25], suggesting that growth arrest but not high calcium is the key signal to turn on K10 expression. Indeed, studies have shown that C/EBPβ expression localizes in the upper differentiated layers of human skin epidermis [26] and C/EBP family of transcription factors inhibit proliferation by blocking cell cycle progression [27, 28]. Therefore it is possible that growth factor depletion condition in culture induces and activates C/EBPs which in turn trigger cell cycle arrest and induction of K10 gene.

K8 and K18 are the keratin pair that is expressed earliest during embryonic development of the epidermis and their expressions are suppressed upon epidermal progenitor cells commitment to keratinocyte lineage [8, 29]. Several transcription factors, such as p63, Ovol2 and Ctip2, have been shown to suppress K8 expression during epidermal development [6, 9, 30]. It has been shown that developing murine p63−/− epidermis expressed high level of K8 and failed to develop a fully mature stratified epidermis [9], suggesting that p63 may regulate epidermal development through suppressing K8 transcription. We have also shown that in wild-type mouse keratinocytes K8 promoter was repressed by transcription factor Ctip2, and upon Ctip2 depletion K8 expression became strongly upregulated in both developing epidermis and in primary mouse keratinocytes at both transcript and protein levels, demonstrating that Ctip2 functions as a transcription suppressor of K8 gene [6].

3. Keratin disorders and skin diseases

3.1. Epidermolysis bullosa simplex diseases and K5/K14

Epidermolysis bullosa simplex (EBS) is an autosomal dominant skin disorder, manifests itself upon trauma in the form of epidermal basal cell death leading to skin blisters [31]. The primary cause of EBS is dominant mutations in either of the genes encoding keratins K5 or K14 [32]. This pair of type I (K14) and type II (K5) is specifically expressed in the basal cell layer of the skin epidermis, which is in direct contact with the basement membrane of the underlying dermis. Therefore, loss of function or mutation of K5 or K14 leads to defects in keratin filament formation in basal keratinocytes and thus triggers skin blistering due to fragility of the basal cell compartment in either human or mouse skin epithelium [33–35] (**Figure 3**). Additionally, it has been shown that overexpression of transcription factor Ovol2 in mouse basal keratinocytes caused reduced K5 and K14 expression, leading to a severe blistering phenotype that resembles the clinical features of EBS [30].

Although K5/K14 mutations are associated with the majority of EBS population, about a quarter of patients with EBS do not have genomic mutations nor abnormal transcript expressions of K5 or K14 [36], suggesting that alternative pathways may be involved in the pathogenesis of these patients. Indeed, recent studies have suggested that in addition to genetic mutation, abnormal K5/K14 functions in basal keratinocytes can also be regulated at post-transcriptional level through post-translational modifications, such as phosphorylation and disulfide bonding. Abnormal phosphorylation of keratins contributes to the pathogenesis and progression of EBS [37]. For example, threonine 150 (T150) in the head domain of K5 was found to be phosphorylated in human EBS keratinocytes. Expression of phosphomimetic T150D K5 mutant in keratinocytes arrested keratin heterodimer assembly leading to impaired

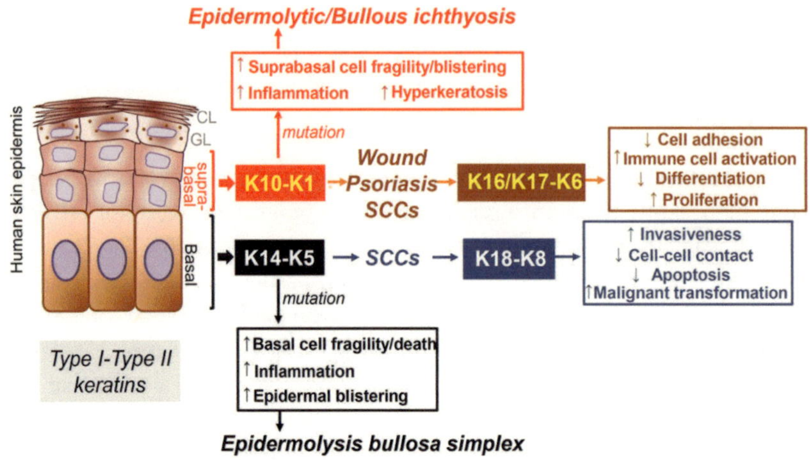

Figure 3. Keratins and skin diseases. Autosomal dominant mutations of basal cell keratins K14 or K5 are associated with skin blistering disease "Epidermolysis Bullosa simplex" and are caused by an increase in basal cell fragility/death, epidermal inflammation and epidermal blistering. Autosomal mutations of the suprabasal KC markers K10 and K1 are associated with hyperkeratosis skin disease "Epidermolytic/Bullous ichthyosis" and are caused by an increase in suprabasal cell fragility/blistering, inflammation, proliferation of basal cells and hyperkeratosis of epithelium. In skin wounds, inflammatory skin diseases such as psoriasis, or squamous cell carcinomas (SCCs), suprabasal cells express high levels of K6, K16 and K17 (keratins associated with activated KC), and these keratins decrease KC adhesion and differentiation and promote immune cell activation and KC proliferation. K8/K18 expression is found in poorly differentiated SCCs with increased invasiveness. K8 or K18 expression in KC leads to decreased cell–cell contact and an increase protection against apoptosis, as well as invasiveness and malignant transformation potential of KC.

keratin filament formations and reduced cell viability and elevated response to stressors [38], suggesting a possible role of K5 T150 phosphorylation in EBS pathogenesis. In addition to phosphorylation, it has been shown that proper disulfide bonding between K14-K5 heterodimers is required to maintain keratin intermediate filament organization and dynamics in primary mouse skin keratinocytes, and disruption of this K14-dependent disulfide linkages may lead to keratinopathies, such as EBS [39].

3.2. Epidermolytic hyperkeratosis diseases and K1/K10

When basal keratinocytes migrate to suprabasal layer of the epidermis, supra basal cells cease transcription of K5 and K14 but instead express K1 and K10. Mutations in either K1 or K10 cause several human skin diseases, such as Epidermolytic ichthyosis (EI), Bullous ichthyosis (BI), palmar-plantar keratoderma and Epidermolytic nevus (EN). BI/EI are caused by rare autosomal dominant mutations of either of K1 or K10 that manifest at birth with fragile blisters and erosions that develop into hyperkeratotic lesions. X-ray crystal structure analysis of K1-K10 heterodimer suggested that point mutation of these keratins may disrupt the disulfide linkage and secondary structure formation between the heterodimer leading to aggregation of the keratin filaments [40]. In contrast to EBI, only the K1 or K10 expressing suprabasal cells are affected in EI/BI patients, and the basal proliferative compartment are not affected. However, these unaffected basal proliferating cells are bathed beneath the rupturing suprabasal cells with inflammatory cytokines, leading to over-proliferation of the basal cells and epithelium – known as hyperkeratosis (**Figure 3**). Therefore, BI skin contains a highly thicken epidermis made up of fragile cells and it is highly susceptible to bacterial and fungal colonization and is highly disfiguring and debilitating for the patient.

In contrast to EI or BI, Epidermolytic nevus (EN) is caused by somatic mutation of either K1 or K10, and the skin blistering phenotype is only affecting part of the body [41, 42]. Because germ line cells are not affected in EN patient, EN parents usually do not transmit the mutations and disease to next generation. However, it has been reported that under rare conditions EN can produce EI in the next generation through transmission from mosaic to germ line. And the risk of disease transmission to the next generation can be evaluated by next generation sequencing of mutation rate in sperm, leukocytes and lesional skin of the EN patients who wish to bear children [43].

3.3. Superficial epidermolytic ichthyosis (SEI) and K2

Superficial epidermolytic ichthyosis (SEI), previously known as ichthyosis bullosa of Siemens (IBS), is an autosomal dominant skin disorder linked to K2 mutations and it is characterized by superficial epidermal fragility and desquamation that lead to characteristic denuded areas. In SEI, aggregates of KF bundles and cytolysis are confined to the upper spinous and granular layers of the epidermis where K2 is expressed [44, 45]. In human, K2 is expressed later in differentiation in the upper spinous and granular layers of skin collected from different body sites [14], but SEI patients usually develop more severe symptoms in palms and soles compared to other body sites suggesting that K2 may play a major role to support tissue integrity in these areas. The tissue specific expression pattern of K2 has been better characterized in mouse skin: in regions of the soles (except foodpad which expresses K2-K9/K10), ears and tail of the mouse, K2 instead of K1 is the major type II keratin that pairs with K10. K2−/-K10−/− double knockout mice or K2−/−mice developed epidermal acanthosis and hyperkeratosis in the tail epidermis, ear epidermis and inter-footpad epidermis of the soles [46, 47], demonstrating that K2-K10 keratin pairs are essential for the epidermal integrity of plantar skin.

3.4. Epidermal palmoplantar keratoderma (EPPK) and K9

Epidermal palmoplantar keratoderma (PPK) is an autosomal dominant skin disorder that develops shortly after birth, and manifests as diffuse hyperkeratosis of the palms and soles and showing sharp demarcations with erythematous margin. Mutations of K9, which is expressed specifically in the suprabasal keratinocytes of the glabrous skin epidermis (palms and soles), are the major cause for EPPK [48, 49]. In human plantar and palmar epidermis, K9 functions as the additional type I keratin, besides K10, to partner with type II keratin K1 [50]. Therefore, mutations of K9 cause pathological epidermal thickening on palms and soles, manifesting as different forms of palmoplantar keratodermas [49]. In mice, K9 expression is restricted to skin epidermis of the footpad, and K9 deficient mice developed calluses marked by hyperpigmentation that are exclusively localized to the stress-bearing footpad. Additionally, hyperproliferation, impaired terminal differentiation, and abnormal expression of K5, K14 and K2 were found in the lesions of K9 deficient mice [51]. Together, these evidence demonstrate that K9 is required for the structural integrity and terminal differentiation of the palmoplantar epidermis.

4. Skin wounds and inflammatory skin diseases and K6/K16/K17

In mammals, the skin of the palm is uniquely adapted to withstand remarkable physical stress, and the palmoplantar epidermis contains a more complex pattern of keratins than thin skin and it is characterized by the constitutive expression of K6, 16, 17 and K9 [52, 53]. Mutations in K6, K16 and K17 genes cause pachyonychia congenita (thick nails, plantar keratoderma) [49].

Distinct from the palmoplantar epidermis, the interfollicular epidermis normally does not express K6, 16 or K17 under homeostatic conditions but these genes can be induced in interfollicular keratinocytes upon activation and reflects a hyper-proliferative state of keratinocytes under wounding or inflammatory conditions [54–56] (**Figure 3**). Upon injury, keratinocytes at the wound edge quickly downregulate K1/K10 and markedly induce K6 (type II)-K16 (type I) keratin heterodimer along with cytoplasmic K17 within 2–6 h of wounding [57], and therefore K6/K16/K17 have been widely used as markers for wound-activated keratinocytes in both human and mouse skin (**Figure 4**). The expressions of K6, 16 and 17 are also elevated in the hyperproliferative epithelium of inflammatory skin diseases such as psoriasis (**Figures 3** and **4**), which shares many inflammatory features with normal skin wounding, including elevated proinflammatory cytokines such as interleukin 1(IL1), tumor necrosis factor α (TNFα), type 1 interferons, interferon γ (IFNγ), IL17 and IL22 [58–60].

Hallmarks of activated keratinocytes include cell hypertrophy, altered cell adhesion and juxtanuclear reorganization of the keratin intermediate filament network, allowing activated keratinocytes to quickly migrate into the wound site, repopulate the skin and restore the epithelial lining and barrier function. In vitro study of mouse keratinocyte overexpressing

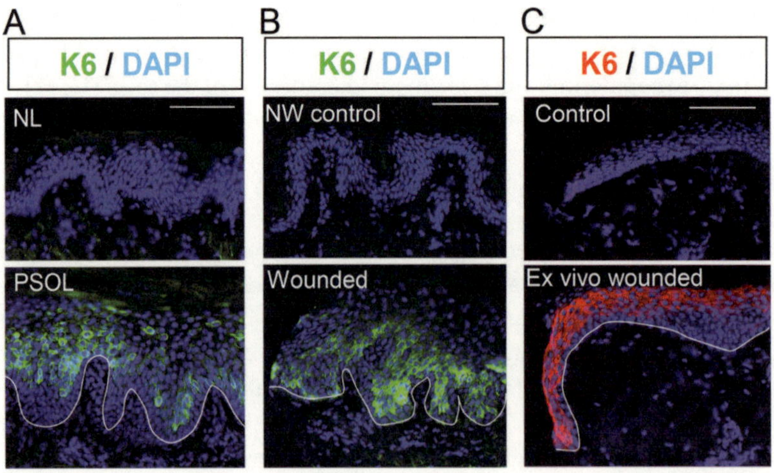

Figure 4. The expression of K6 is elevated in both wounded and psoriatic human skin epidermis. Human skin sections from (A) psoriasis lesional (PSOL), (B) in vivo wounded skin (by punch biopsy and collected at day3), or (C) ex vivo wounded skin were stained with K6 antibody in green or red as indicated. Non-wounded nuclei were counterstained with DAPI (blue). Scale bar, 100 μm. In A, non-lesional skin (NL) from the same patient was used as control for PSO. In B, skin biopsy collected at day 0 was used as non-wounded control. Note that while K6 expression was not detected in all control skin epidermis, strong K6 staining was detected in the suprabasal layers of both wounded and psoriatic skin epidermis in similar patterns. Also K6 was strongly elevated in the migrating tongue of ex vivo wounded human skin at the wound edge. Details of these samples can be found in our previous published work [58].

K16 revealed that while forced expression of K16 did not alter cell proliferation, it caused a reduction in cell adhesion and K10 expression (early differentiation) [61]. K10 expression inhibits cell proliferation, but ectopic expression of K16 promotes cell proliferation and diminishes the inhibitory function of K10 on cell proliferation when K6 and K10 are co-expressed [11] . K6−/− mouse keratinocytes migrated faster than control wild-type cells [56], and K6 negatively regulates the migratory potential of skin keratinocytes by inhibiting Src kinase [62], suggesting that the migratory feature of activated keratinocytes may be regulated by an K6/K16 independent pathway or by a non-cell autonomous manner. In contrast to K6−/− keratinocytes, K17−/− mouse keratinocytes show a delay in the closure of wounds [63], and protein translation, AKT activity and cell proliferation are suppressed in K17−/− keratinocytes [64]. These results suggest that in activated keratinocytes while the cell adhesion and differentiation maybe regulated by K6, cell hyper-proliferation and migration in response to wound maybe controlled by K17.

4.1. Regulation of K6/16/17 expression

Our group has shown that damage-associated molecular patterns "DAMPs", such as double-stranded RNA, are the first signals released from necrotic cells to rapidly initiate the inflammatory cytokine production from surrounding undamaged human keratinocyte upon injury [58, 65]. Inflammatory cytokines released either from activated keratinocytes or from recruited immune cells initiate expression of hyperproliferative keratins and keratinocyte hyperproliferation. It has been shown that cytokines interleukin 1(IL1) and tumor necrosis factor α (TNFα) can synergistically induce the transcription of K6 through transcription factor C/EBPβ and NFκB [66]. Transcription factor AP1 (c-Fos + c-Jun) can also activate K6 promoter synergistically with NFκB under inflammatory conditions [19]. Additionally, transcription factor NRF2 translocated from cytoplasm to nucleus upon stimulation with proinflammatory cytokines such as IL 17 or IL 22, and upregulated the expression of K6, K16 and K17 genes via the antioxidant responsive element (ARE)-binding region, promoting proliferation of keratinocytes in psoriasis [67]. K17 expression could also be induced by IFNγ in skin epidermis, and the K17 in turn function as an auto-antigen to stimulate proliferation of T cells and IFN expression, contributing to the amplification of the autoimmune response and immunopathogenesis of psoriasis [68, 69]. Accumulating recent evidences have suggested that microRNAs also play important roles in modulating keratinocyte activation and skin inflammation [70, 71]. Keratin expression can be regulated by microRNA in the contact of psoriasis. For example, miR-486-3p targeted K17 mRNA for degradation and it was identified as the top downregulated microRNAs in psoriasis, leading to K17 overproduction and hyperproliferation in psoriatic keratinocytes [72].

5. Keratins and cutaneous squamous cell carcinomas

5.1. SCCs and K5/K14/K15 and K6/K16/17

Cutaneous squamous cell carcinoma (SCC) is the second most common skin cancers and represents about 20% of all skin cancer, with up to 700,000 new cases annually diagnosed in the USA, and it is associated with a substantial risk of metastasis [73]. SCC is characterized by extensive expression of K5/K14 through the epidermis and the expression of hyperproliferative keratin K6, K16 and K17 [50], which are not only upregulated in inflammatory skin, but often upregulated in many tumors originating in stratified and pseudostratified epithelia. K1/K10 may also be focally expressed in SCCs, and K8/K18 is often detected in poorly differentiated SCCs, and the role of K8/K18 will be discussed in more details next.

5.2. SCCs and K8/K18

As we have described in chapter 2, the simple epithelia-specific keratin pair, K8/K18, are expressed in keratinocyte progenitors, early on during embryonic skin development, and upon vertical epidermal stratification K8/K18 expression is then substituted by K5/K14, and becomes eventually lost in fully mature skin epidermis [9, 10, 74]. Overexpressing human K8 in mouse epidermis (TGHK8 mice) lead to severe epidermal phenotypes including epidermal hyperplasia associated with orthokeratotic hyperkeratosis, dysplastic hair follicles and altered expression terminal differentiation markers [75]. The severity of these skin phenotypes increased during aging, and the aged TGHK8 mice developed spontaneous premalignant skin tumors, and TGHK8 mice showed a drastic increase in the malignant progression of skin tumors in mouse model of chemical skin carcinogenesis. Previously, we have shown that Ctip2−/− mouse keratinocytes that overexpressed K8 exhibited loss of cell–cell contact and contain much thicker central stress fibers [6], indicating that aberrant K8 expression may decrease cell–cell adhesion and trigger an EMT (epithelial–mesenchymal transition) phenotype in epidermal keratinocytes. These results suggest that expression of K8 in adult epidermis impairs the normal epidermal differentiation program and may be responsible for the invasive behavior of transformed epidermal keratinocytes. Indeed, in adult skin epidermis, aberrant K8/K18 expression is broadly correlated with increased invasiveness and poor prognosis of squamous cell carcinomas [76]. In addition, K8/K18 protect epithelial cells against apoptosis mediated by proapoptotic signals, such as TNFα [77] released by macrophages and T lymphocytes and Fas [78], and may enable cancer cells to resist immune cell-mediated cell death and escape immune surveillance.

5.3. Regulation of keratin expression and function in SCCs

Phosphorylation is the key post-translational modification that regulates keratin functions, and phosphorylation of K8 is among the most well studied in keratin family [79]. More than a dozen phosphorylation sites have been identified on serine residues of K8 [79], and phosphorylation of K8 enhances the migratory, proliferative and invasive potential of epithelial cells, therefore promoting the malignant transformation of cancer cells [80]. Keratin expression can be altered through epigenetic modification during malignant transformation. A recent study investigating genome-wide DNA methylation changes in the progression from healthy human epidermis to cSCC reveals that DNA methylation profiles of cSCC epidermis display classical features of cancer methylomes compared to normal epidermis samples [81]. Further analyses of DNA methylation patterns of keratin gene clusters (including basal cell keratins K5 and K14 which are ectopically expressed throughout cSCC epidermis) identified major DNA methylation differences between healthy donors and cSCC patients, suggesting that abnormal keratin expression in SCCs may be regulated through epigenetic mechanism such as DNA methylation.

6. Conclusion

Tissue and cell differentiation specific expression of pair between type I and type II Keratins play essential roles in forming the intermediate filaments and providing cytoskeletal and structural support and mechanical resilience for epithelia tissues. In addition to these structural roles, keratins also control cell migration, cell adhesion, proliferation and differentiation processes in keratinocytes and modulate immune system in various settings. The transcriptions of keratins are tightly controlled by a series of transcription factors, and keratin functions are also regulated by post-translational modifications such as phosphorylation and intermolecular disulfide bond formation. Mutation or dysfunction of basal keratins K14/K5 or suprabasal keratins K1/K10 lead to severe skin blistering diseases, whereas K6, K16 and K17 are rapidly induced in activated keratinocytes upon skin wounding and are also expressed in inflammatory skin diseases, such as psoriasis. Understanding keratin functions and related regulatory mechanisms will help to design new therapeutic interventions for keratin-related skin diseases.

Acknowledgements

This work was supported by National Institute of Arthritis and Musculoskeletal and Skin Diseases grant (R01AR069653).

Conflict of interest

The author has nothing to disclose.

Author details

Ling-juan Zhang[1,2]

Address all correspondence to: lingjuan.zhang@xmu.edu.cn

1 School of Pharmaceutical Sciences, Xiamen University, Xiamen, China

2 Department of Dermatology, School of Medicine, UC San Diego, USA

References

[1] Franke WW et al. Diversity of cytokeratins - differentiation specific expression of cytokeratin polypeptides in epithelial-cells and tissues. Journal of Molecular Biology. 1981;**153**(4):933-959

[2] Moll R et al. The catalog of human cytokeratins - patterns of expression in normal epithelia, tumors and cultured-cells. Cell. 1982;**31**(1):11-24

[3] Schweizer J et al. New consensus nomenclature for mammalian keratins. The Journal of Cell Biology. 2006;**174**(2):169-174

[4] Herrmann H et al. Intermediate filaments: Primary determinants of cell architecture and plasticity. Journal of Clinical Investigation. 2009;**119**(7):1772-1783

[5] Sambandam SAT et al. 14-3-3 sigma regulates keratinocyte proliferation and differentiation by modulating Yap1 cellular localization. Journal of Investigative Dermatology. 2015; **135**(6):1621-1628

[6] Zhang LJ et al. Ctip2 is a dynamic regulator of epidermal proliferation and differentiation by integrating EGFR and notch signaling. Journal of Cell Science. 2012;**125**(23):5733-5744

[7] Fuchs E. Scratching the surface of skin development. Nature. 2007;**445**(7130):834-842

[8] Nagarajan P, Romano RA, Sinha S. Transcriptional control of the differentiation program of interfollicular epidermal keratinocytes. Critical Reviews in Eukaryotic Gene Expression. 2008;**18**(1):57-79

[9] Romano RA et al. Delta Np63 knockout mice reveal its indispensable role as a master regulator of epithelial development and differentiation. Development. 2012;**139**(4):772-782

[10] Liu S, Zhang H, Duan E. Epidermal development in mammals: Key regulators, signals from beneath, and stem cells. International Journal of Molecular Sciences. 2013;**14**(6): 10869-10895

[11] Paramio JM et al. Modulation of cell proliferation by cytokeratins K10 and K16. Molecular and Cellular Biology. 1999;**19**(4):3086-3094

[12] Chen JL et al. An unexpected role for keratin 10 end domains in susceptibility to skin cancer. Journal of Cell Science. 2006;**119**(24):5067-5076

[13] Bose A et al. Keratin K15 as a biomarker of epidermal stem cells. International Journal of Molecular Sciences. 2013;**14**(10):19385-19398

[14] Collin C et al. Characterization of human cytokeratin 2, an epidermal cytoskeletal protein synthesized late during differentiation. Experimental Cell Research. 1992;**202**(1):132-141

[15] Lloyd C et al. The basal keratin network of stratified squamous epithelia: Defining K15 function in the absence of K14. The Journal of Cell Biology. 1995;**129**(5):1329-1344

[16] Lyle S et al. The C8/144B monoclonal antibody recognizes cytokeratin 15 and defines the location of human hair follicle stem cells. Journal of Cell Science. 1998;**111**(Pt 21):3179-3188

[17] Liu Y et al. Keratin 15 promoter targets putative epithelial stem cells in the hair follicle bulge. The Journal of Investigative Dermatology. 2003;**121**(5):963-968

[18] Rossi A et al. Effect of AP1 transcription factors on the regulation of transcription in normal human epidermal keratinocytes. The Journal of Investigative Dermatology. 1998; **110**(1):34-40

[19] Ma S et al. Transcriptional control of K5, K6, K14, and K17 keratin genes by AP-1 and NF-kappaB family members. Gene Expression. 1997;**6**(6):361-370

[20] Tomic-Canic M et al. Novel regulation of keratin gene expression by thyroid hormone and retinoid receptors. The Journal of Biological Chemistry. 1996;**271**(3):1416-1423

[21] Tomic-Canic M et al. Identification of the retinoic acid and thyroid hormone receptor-responsive element in the human K14 keratin gene. The Journal of Investigative Dermatology. 1992;**99**(6):842-847

[22] Perez P et al. Altered skin development and impaired proliferative and inflammatory responses in transgenic mice overexpressing the glucocorticoid receptor. The FASEB Journal. 2001;**15**(11):2030-2032

[23] Zhu S et al. C/EBPbeta modulates the early events of keratinocyte differentiation involving growth arrest and keratin 1 and keratin 10 expression. Molecular and Cellular Biology. 1999;**19**(10):7181-7190

[24] Maytin EV et al. Keratin 10 gene expression during differentiation of mouse epidermis requires transcription factors C/EBP and AP-2. Developmental Biology. 1999;**216**(1):164-181

[25] Li F, Adase CA, Zhang LJ. Isolation and culture of primary mouse keratinocytes from neonatal and adult mouse skin. Journal of Visualized Experiments. 2017;**125**:1-6, e56027

[26] Sikder H et al. A central role for transcription factor C/EBP-beta in regulating CD1d gene expression in human keratinocytes. Journal of Immunology. 2009;**183**(3):1657-1666

[27] Johnson PF. Molecular stop signs: Regulation of cell-cycle arrest by C/EBP transcription factors. Journal of Cell Science. 2005;**118**(Pt 12):2545-2555

[28] Timchenko NA et al. CCAAT/enhancer-binding protein alpha (C/EBP alpha) inhibits cell proliferation through the p21 (WAF-1/CIP-1/SDI-1) protein. Genes & Development. 1996; **10**(7):804-815

[29] Oshima RG et al. Intermediate filament protein synthesis in preimplantation murine embryos. Developmental Biology. 1983;**99**(2):447-455

[30] Lee B et al. Overexpression of transcription factor Ovol2 in epidermal progenitor cells results in skin blistering. The Journal of Investigative Dermatology. 2017;**137**(8):1805-1808

[31] Coulombe PA. Discovery of keratin function and role in genetic diseases: The year that 1991 was. Molecular Biology of the Cell. 2016;**27**(18):2807-2810

[32] Bonifas JM, Rothman AL, Epstein EH, Epidermolysis bullosa simplex: Evidence in two families for keratin gene abnormalities. Science. 1991;**254**(5035):1202-1205

[33] Coulombe PA, Lee CH. Defining keratin protein function in skin epithelia: Epidermolysis bullosa simplex and its aftermath. The Journal of Investigative Dermatology. 2012;**132**(3 Pt 2): 763-775

[34] Coulombe PA et al. Point mutations in human keratin 14 genes of epidermolysis bullosa simplex patients: Genetic and functional analyses. Cell. 1991;**66**(6):1301-1311

[35] Vassar R et al. Mutant keratin expression in transgenic mice causes marked abnormalities resembling a human genetic skin-disease. Cell. 1991;**64**(2):365-380

[36] Bolling MC et al. Mutations in KRT5 and KRT14 cause epidermolysis bullosa simplex in 75% of the patients. British Journal of Dermatology. 2011;**164**(3):637-644

[37] Sawant MS, Leube RE. Consequences of keratin phosphorylation for cytoskeletal organization and epithelial functions. International Review of Cell and Molecular Biology. 2017; **330**:171-225

[38] Sawant M et al. Threonine 150 phosphorylation of keratin 5 is linked to Epidermolysis Bullosa simplex and regulates filament assembly and cell viability. The Journal of Investigative Dermatology. 2018;**138**(3):627-636

[39] Feng X, Coulombe PA. A role for disulfide bonding in keratin intermediate filament organization and dynamics in skin keratinocytes. Journal of Cell Biology. 2015;**209**(1):59-72

[40] Bunick CG, Milstone LM. The X-ray crystal structure of the keratin 1-keratin 10 Helix 2B heterodimer reveals molecular surface properties and biochemical insights into human skin disease. The Journal of Investigative Dermatology. 2017;**137**(1):142-150

[41] Tsubota A et al. Keratin 1 gene mutation detected in epidermal nevus with epidermolytic hyperkeratosis. The Journal of Investigative Dermatology. 2007;**127**(6):1371-1374

[42] Paller AS et al. Genetic and clinical mosaicism in a type of epidermal nevus. The New England Journal of Medicine. 1994;**331**(21):1408-1415

[43] Kono M et al. A child with epidermolytic ichthyosis from a parent with epidermolytic nevus: Risk evaluation of transmission from mosaic to Germline. The Journal of Investigative Dermatology. 2017;**137**(9):2024-2026

[44] Rothnagel JA et al. Mutations in the rod domain of keratin 2e in patients with ichthyosis bullosa of Siemens. Nature Genetics. 1994;**7**(4):485-490

[45] Tsubota A et al. Congenital ichthyosiform erythroderma mimicking ichthyosis bullosa of Siemens. The British Journal of Dermatology. 2008;**158**(1):191-194

[46] Fischer H et al. Keratins K2 and K10 are essential for the epidermal integrity of plantar skin. Journal of Dermatological Science. 2016;**81**(1):10-16

[47] Fischer H et al. Loss of keratin K2 expression causes aberrant aggregation of K10, hyperkeratosis, and inflammation. The Journal of Investigative Dermatology. 2014;**134**(10):2579-2588

[48] Langbein L et al. Molecular characterization of the body site-specific human epidermal cytokeratin 9: cDNA cloning, amino acid sequence, and tissue specificity of gene expression. Differentiation. 1993;**55**(1):57-71

[49] Toivola DM et al. Keratins in health and disease. Current Opinion in Cell Biology. 2015;**32**: 73-81

[50] Moll R, Divo M, Langbein L. The human keratins: Biology and pathology. Histochemistry and Cell Biology. 2008;**129**(6):705-733

[51] Fu DJ et al. Keratin 9 is required for the structural integrity and terminal differentiation of the Palmoplantar epidermis. Journal of Investigative Dermatology. 2014;**134**(3):754-763

[52] Swensson O et al. Specialized keratin expression pattern in human ridged skin as an adaptation to high physical stress. British Journal of Dermatology. 1998;**139**(5):767-775

[53] Maruthappu T et al. Rhomboid family member 2 regulates cytoskeletal stress-associated keratin 16. Nature Communications. 2017;**8**:1-11, 14174

[54] Paladini RD et al. Onset of re-epithelialization after skin injury correlates with a reorganization of keratin filaments in wound edge keratinocytes: Defining a potential role for keratin 16. The Journal of Cell Biology. 1996;**132**(3):381-397

[55] Leigh IM et al. Keratins (K16 and K17) as markers of keratinocyte hyperproliferation in psoriasis in vivo and in vitro. The British Journal of Dermatology. 1995;**133**(4):501-511

[56] Wong P, Coulombe PA. Loss of keratin 6 (K6) proteins reveals a function for intermediate filaments during wound repair. The Journal of Cell Biology. 2003;**163**(2):327-337

[57] McGowan K, Coulombe PA. The wound repair-associated keratins 6, 16, and 17. Insights into the role of intermediate filaments in specifying keratinocyte cytoarchitecture. Sub-Cellular Biochemistry. 1998;**31**:173-204

[58] Zhang LJ et al. Antimicrobial peptide LL37 and MAVS signaling drive interferon-beta production by epidermal keratinocytes during skin injury. Immunity. 2016;**45**(1):119-130

[59] Nickoloff BJ et al. Lessons learned from psoriatic plaques concerning mechanisms of tissue repair, remodeling, and inflammation. Journal of Investigative Dermatology Symposium Proceedings. 2006;**11**(1):16-29

[60] Tampa M et al. The pathophysiological mechanisms and the quest for biomarkers in psoriasis, a stress-related skin disease. Disease Markers. 2018;**2018**:5823684

[61] Wawersik M, Coulombe PA. Forced expression of keratin 16 alters the adhesion, differentiation, and migration of mouse skin keratinocytes. Molecular Biology of the Cell. 2000; **11**(10):3315-3327

[62] Rotty JD, Coulombe PA. A wound-induced keratin inhibits Src activity during keratinocyte migration and tissue repair. The Journal of Cell Biology. 2012;**197**(3):381-389

[63] McGowan KM et al. Keratin 17 null mice exhibit age- and strain-dependent alopecia. Genes & Development. 2002;**16**(11):1412-1422

[64] Kim S, Wong P, Coulombe PA. A keratin cytoskeletal protein regulates protein synthesis and epithelial cell growth. Nature. 2006;**441**(7091):362-365

[65] Lai Y et al. Commensal bacteria regulate toll-like receptor 3-dependent inflammation after skin injury. Nature Medicine. 2009;**15**(12):1377-1382

[66] Komine M et al. Interleukin-1 induces transcription of keratin K6 in human epidermal keratinocytes. The Journal of Investigative Dermatology. 2001;**116**(2):330-338

[67] Yang L et al. Nrf2 promotes keratinocyte proliferation in psoriasis through up-regulation of keratin 6, keratin 16, and keratin 17. The Journal of Investigative Dermatology. 2017; **137**(10):2168-2176

[68] Gudmundsdottir AS et al. Is an epitope on keratin 17 a major target for autoreactive T lymphocytes in psoriasis? Clinical and Experimental Immunology. 1999;**117**(3):580-586

[69] Shen Z et al. Altered keratin 17 peptide ligands inhibit in vitro proliferation of keratinocytes and T cells isolated from patients with psoriasis. Journal of the American Academy of Dermatology. 2006;**54**(6):992-1002

[70] Hawkes JE et al. microRNAs in psoriasis. The Journal of Investigative Dermatology. 2016; **136**(2):365-371

[71] Hermann H et al. miR-146b probably assists miRNA-146a in the suppression of keratinocyte proliferation and inflammatory responses in psoriasis. The Journal of Investigative Dermatology. 2017;**137**(9):1945-1954

[72] Jiang M et al. TGFbeta/SMAD/microRNA-486-3p signaling Axis mediates keratin 17 expression and keratinocyte Hyperproliferation in psoriasis. The Journal of Investigative Dermatology. 2017;**137**(10):2177-2186

[73] Alam M, Ratner D. Cutaneous squamous-cell carcinoma. The New England Journal of Medicine. 2001;**344**(13):975-983

[74] Zhao Q et al. Spatiotemporal expression of p63 in mouse epidermal commitment. International Journal of Molecular Sciences. 2015;**16**(12):29542-29553

[75] Casanova ML et al. Epidermal abnormalities and increased malignancy of skin tumors in human epidermal keratin 8-expressing transgenic mice. The FASEB Journal. 2004;**18**(13): 1556-1558

[76] Karantza V. Keratins in health and cancer: More than mere epithelial cell markers. Oncogene. 2011;**30**(2):127-138

[77] Caulin C et al. Keratin-dependent, epithelial resistance to tumor necrosis factor-induced apoptosis. The Journal of Cell Biology. 2000;**149**(1):17-22

[78] Gilbert S et al. Simple epithelium keratins 8 and 18 provide resistance to Fas-mediated apoptosis. The protection occurs through a receptor-targeting modulation. The Journal of Cell Biology. 2001;**154**(4):763-773

[79] Kim HJ, Choi WJ, Lee CH. Phosphorylation and reorganization of keratin networks: Implications for carcinogenesis and epithelial mesenchymal transition. Biomolecules & Therapeutics (Seoul). 2015;**23**(4):301-312

[80] Tiwari R et al. Quantitative phosphoproteomic analysis reveals system-wide signaling pathways regulated by site-specific phosphorylation of Keratin-8 in skin squamous cell carcinoma derived cell line. Proteomics. 2017;**17**(7):1-17, 1600254

[81] Rodriguez-Paredes M et al. Methylation profiling identifies two subclasses of squamous cell carcinoma related to distinct cells of origin. Nature Communications. 2018;**9**(1):1-9, 577

Section 3

Physical and Chemical Characteristics

Chapter 6

Shape Memory Effect of Keratin Fibers

Xueliang Xiao and Jin-lian Hu

Additional information is available at the end of the chapter

http://dx.doi.org/10.5772/intechopen.77248

Abstract

In the past thousands of years, keratin fibers were only considered as textile fibers for excellent fiber performances, such as high strength, acceptable elasticity, good thermal insulation, etc. Only recently, some indications have been obtained that keratin fiber may be a smart natural material that may subvert people's perception of this matter. The smart attribute displays shape memory effects (SMEs) responsive to many types of stimuli including water, heat, coupled water-heat, redox agents, UV light, etc. These smart functions of keratin fibers are found to be the result of three structural components: crystals, hydrogen bonds (HBs), and disulfide bonds (DBs) among intra- and inter-keratin macromolecules. In this chapter, keratin fibers (such as camel hair) were employed for investigating their SMEs under five types of stimuli, in which the HBs, DBs, and crystals were characterized separately, as well as the fiber shape fixation and recovery ratios, respectively. The whole test results indicated that keratin hair fiber is a type of shape memory polymer and the related SME depends on the contents of the HBs, DBs, and crystalline phase inside the hair.

Keywords: keratin fiber, shape memory effect, stimulus-responsive behavior

1. Introduction

Shape memory polymers (SMPs) are a kind of smart polymers from the viewpoint of their stimuli-responsive behavior adapting our human demands. SMPs have a general net-point-switch structure [1] where net-points determine the permanent shape and reversible switch leads to the temporary shape. Most SMPs show one stimulus in the current existing cases although there are a number of recent reports of triple stimuli of SMPs [2–5]. Nowadays, people begin to realize the importance of developing multi-stimuli-responsive SMPs; for example, Schattling [6] addressed the rising importance of multi-stimuli polymers that are, namely, "all-in-one talents" in many areas, like comparable natural adaptability in life sciences and the parallel writing of information to give a dramatic increase of memory density in information technology. The authors believed that a polymer with one more stimulus can show a higher degree of control for more free choice and higher level of intricacy, in turn to improve the adaptability of the polymer in diverse environments with improved smart functions. Nevertheless, the development of such high intelligent polymers requires novel sciences and technologies that might go beyond our general imagination. Conversely, wisdom is encountered anywhere in nature, and many tricks can be learned even from a subject that is fairly common like a keratin animal hair [7]. For instance, a dry woolen yarn shows an interesting shape memory effect (SME) when it is immersed in water with reported 20% contraction after processes of being treated by $NaHSO_3$ dilute solution and pre-stretched with 20–30% of strain [8]. The contraction phenomenon is believed to be a dual stimulus SME of woolen yarn responsive to water and redox.

For thousands of years, animal hair (a kind of keratin fiber) has been merely considered as a textile fiber with outstanding performances for its excellent elasticity and thermal insulation. These outstanding properties are ascribed to the keratin hierarchical structure with macro- and microfibrils and helical coils, being wrapped by cortex and cuticles at the outside of the hair as shown in **Figure 1a**. In addition, the structural components of hydrogen (HB) and disulfide (DB) bonds between macromolecules in crystals and amorphous regions play key roles in the properties, as shown schematically in **Figure 1b**. HB is based on electrostatic attraction between opposite polar molecules, namely, dipole–dipole attraction, which in fact is not a true bond. Conversely, DB as a covalent cross-link for keratin molecules (can be characterized by Raman spectroscopy) [10] determines the elasticity of the hair [11, 12]. The HBs and DBs in

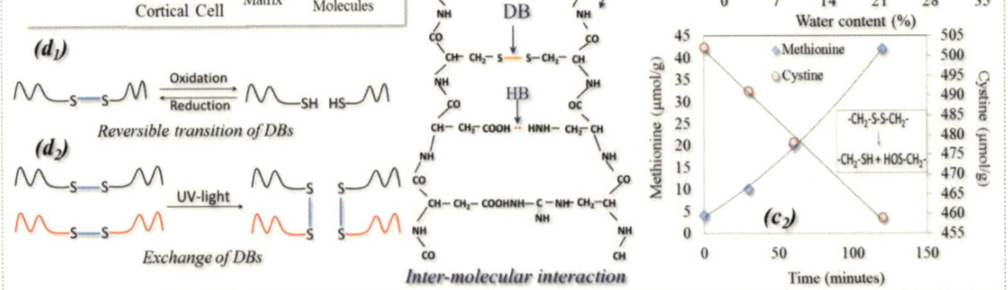

Figure 1. Components and properties of a wool fiber: (a) wool hierarchical structure from the cuticle to helical coils; (b) HBs, DBs, and crystals in wool macromolecules; (c_1) water content influencing wool glass transition temperature; (c_2) reducing reaction influencing amount of DBs. Schematic illustration of reversible reaction of DBs: (d_1) in redox reaction and (d_2) under UV light radiation [9].

hair amorphous regions were reported reversible under certain conditions, such as processed with water, heat, and redox agents [13, 14], respectively; these materials can result in conversion of keratin fiber with temporary and permanent shapes macroscopically [15]. Moreover, in **Figure 1b**, the schematic illustration showed the crystallinity of wool (around 40%) difficult to be cleaved, and the crystalline phase was studied by some textile scholars using X-ray diffraction (XRD) technique that such phase is composed of dense regular HBs [16–18].

Regarding the HBs acting as dense physical cross-linking in polymers, they perform as hard polymer crystal phase that is difficult to cleave. Such cross-linking is well known in SMPs, and the regular dense HBs are net-points in many water- and UV-induced SMPs, in which HB-based crystal acts as stress-stored net-point for shape recovery from deformation [19, 20]. When the cross-linking is not strong enough that may be reversible between decomposition and re-formation under certain reversal conditions like heat and moisture [21–23], this kind of cross-linking can be used as switch unit in SMPs. As to the keratin fiber, for example, the glass transition temperature (T_g) of wool, was found to decrease when its water content is increased (**Figure 1c$_1$**) [24], this is because of a large amount of HBs that are collapsed in the amorphous area under the penetrated aqueous molecules. In turn, the collapsed HBs were also observed to re-form in the amorphous area of wool when the water content is decreased under drying condition. Similarly, such kind of HBs performs reversible behavior with and without water alternatively [25]. Therefore, it can be confirmed that HBs act as a key role in water-/moisture-driven shape memory behavior [7, 22, 26]. In addition, a poly(ethylene glycol) (PEG)-based polyurethane was observed with a water-induced SME, where regular dense HBs were found to be dissolved by aqueous molecules due to the collapse of HBs [27]. When the molecular weight of PEG is increased, a copolymer with PEG molecules was synthesized with a shape recovery rate of 99% using HB as switch [28]. A simple composite with controllable triple SME supramolecular was also developed successfully using HB as switch between macromolecule branches and mesogenic units [29] in which the HBs can improve the broad and independent control of T_g and density of cross-linking [30].

For covalent bond of DB, it can be collapsed either when a reductant agent is used to react with such bond, and this can be reflected through the amount of cysteine and methionine with opposite tendency during the reducing reaction, as shown in **Figure 1c$_2$**[31]. Here, the covalent bond of DB can be broken into two thiol groups after the reducing reaction. In turn, the two thiol groups can be reversible to regenerate a DB after an oxidation reaction, such as in a H_2O_2 solution [32]. DB, in this case, performs a kind of dynamic bond like HB responsive to redox agents or high-energy light triggers that can act as switch role in some SMPs. **Figure 1d$_1$**shows the reversible behavior taking place between thiol groups and DBs [33]. The reversible behavior was reported in many responsive polymer capsules, micelles, and gels for many applications like drug delivery or self-healing [34–36]. As to the self-healing, DB shows another applied type of reversibility. A specific polymer with DB can be driven for self-healing by triggers like UV radiation for exchange reaction, and an increase of DBs shows an increased healing ability, as shown in **Figure 1d$_2$** [37–40]. So far, two references were found, one thermal-responsive SMP using semicrystalline as net-point and DBs for covalently network were used to cross-link for self-healing under UV light [41]. In another reference, a SMP with DBs was reported using DB as switch for SME when responsive to redox, cellulose derivatives in the polymer were used with cross-linked mercapto groups [42]. Strictly speaking, the report of using DBs as net-points in SMPs has not been found yet; however, the covalent bonds are strong enough to be stable under certain conditions that manifest high opportunity to be net-points in SMPs such as in wool hair. The DBs in wool as net-points can control the wool's elasticity significantly under general environmental factors [11, 12]. In a conclusion, DBs as reversible component could be switches responsive to redox, showing self-healing capability in some polymers, when reversible reactions take place between keratin macromolecules under specific (UV radiation) conditions [43].

So far, the three components in polymer (crystalline phase, HBs, and DBs) may form or tailor many types of combinations for a variety of SMEs and performances [44]. It is interesting to develop a kind of multi-responsive SMPs through studying and mimicking the components, structures, and related shape memory functions of keratin fiber. This would inevitably enrich the super intelligent family of SMPs.

2. Water-sensitive shape memory behavior of keratin fiber

As is known, a dry mature animal hair usually has three layers, i.e., a porous medulla in the hair center, a middle layer of cortex, and a surface layer of tiles-overlapping scales (around 0.5 μm of thickness). In this section, as shown in the SEM images in **Figure 2a**, three kinds of keratin hairs evidence the three-layer structure under a dry state. The volume ratio of the cortex and the texture of the medulla result in a bit difference of the configuration of three keratin fibers.

The central medulla is a form of porous medium made of biodegradable cellular materials. The porous structure benefits the warmth retention of keratin hair due to the stored still air. The middle layer of hair cortex accounts for the main volume of the hair and provides the main strength to deformation when facing external force. A two-phase model was once proposed to interpret the hair context that employs crystalline phase and matrix (two-phase) to simplify the hair [45, 46]. In water, the hair cortex swells evidently, as shown in **Figure 2a**. This means that the aqueous molecules can penetrate easily into the matrix phase of the cortex, resulting in the thicker of wet hair with the narrower of the central medulla in comparison with dry hair. Here, a large amount of reversible polar groups in amorphous regions of the hair can interact with penetrated aqueous molecules based on the two-phase model. The penetration and interaction of aqueous molecules with HBs can enlarge the space of neighboring polypeptide chains by breaking HBs between groups of N—H and C=O at adjacent polypeptide branches. The cortex volume is observed to increase with diametral swelling from the macroscopic point of view, which relates to the quasi-plastic distortion by means of changing the matrix phase of weak HBs between macromolecule branches.

With respect to shape memory behavior, as shown in **Figure 2b**, a mature camel guard hair shows a straight smooth configuration along its axis. The innate shape was deformed manually with an entanglement of the camel hair onto a circular plastic bar and immersion of the deformed hair into the cooled water. After a certain period of deformed hair in water, it is believed that the hair was plasticized and wrapped temporarily on the bar. A drying process gives rise to the hair in helical profile under a free state. A recovery process was given to the

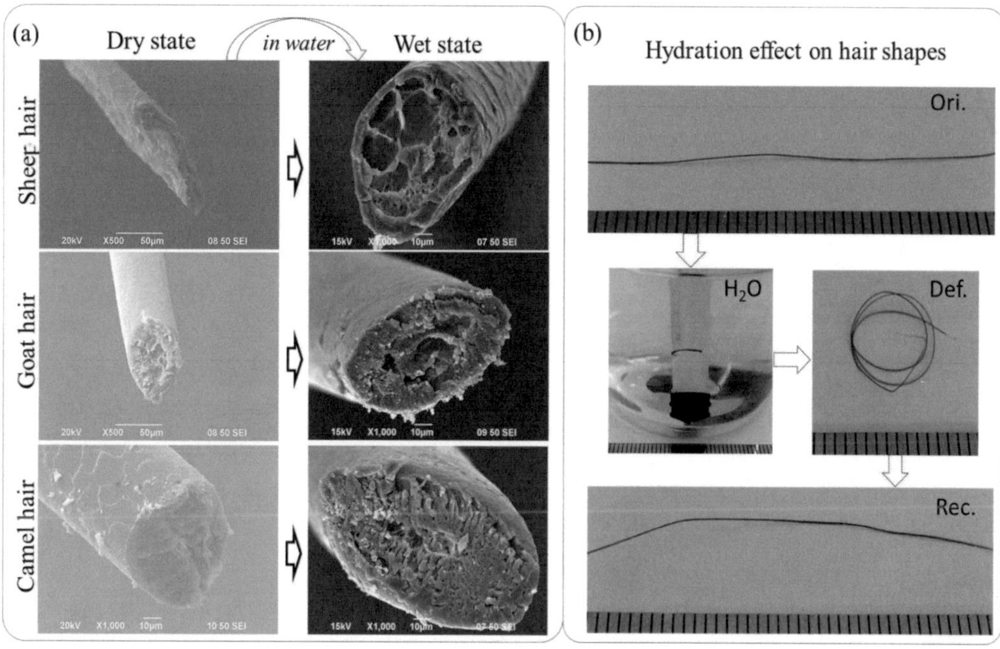

Figure 2. Hydration of three animal hairs: (a) cross sections of dry and wet sheep, goat, and camel hairs in SEM images and (b) demonstration of the water-sensitive shape memory behavior of a typical keratin fiber, camel guard hair, at the innate shape, temporarily fixed shape after hydration, and recovered shape after in water. Note: All figures were obtained at 20°C [7].

deformed configuration that the spatial structure recovers back to a straight shape soon after the deformed hair was put in water again. The recovery process and related dynamics of the deformed hair in water can refer to the author's work [7]. In comparison, the recovered camel hair and innate hair reveal almost the same shape without any spatial residues.

Figure 3 shows a quantification measure of water-stimulated SME of keratin fibers, in which the hair SME was investigated using the variation of folded hair in triangle into a certain angle (θ_s) under the transition of hair states (dry↔wet). Here, a perfect SME of hair displays the variation of deformed angles of "$\theta = 180° \rightarrow \theta = \theta_s° \rightarrow \theta = 180°$" for the aqueous molecules to penetrate into and remove out of hairs. Nevertheless, the ideal (100%) shape fixation and recovery cannot be realized for most natural keratin fibers, because the fibers show non-consistent network and phases between net-points. Therefore, two variables are used to justify the SME of keratin hair, i.e., $\theta = \theta_f$ and $\theta = \theta_r$, standing for the fixed shape angle and recovered shape angle, respectively. The shape fixation ratio (R_f, Eq. (1)) and shape recovery ratio (R_r, Eq. (2)) [47] are thereafter derived on the basis of the two measured angles:

$$R_f = \frac{180 - \theta_f}{180 - \theta_s} \quad (1)$$

$$R_r = \frac{\theta_r - \theta_s}{180 - \theta_s} \quad (2)$$

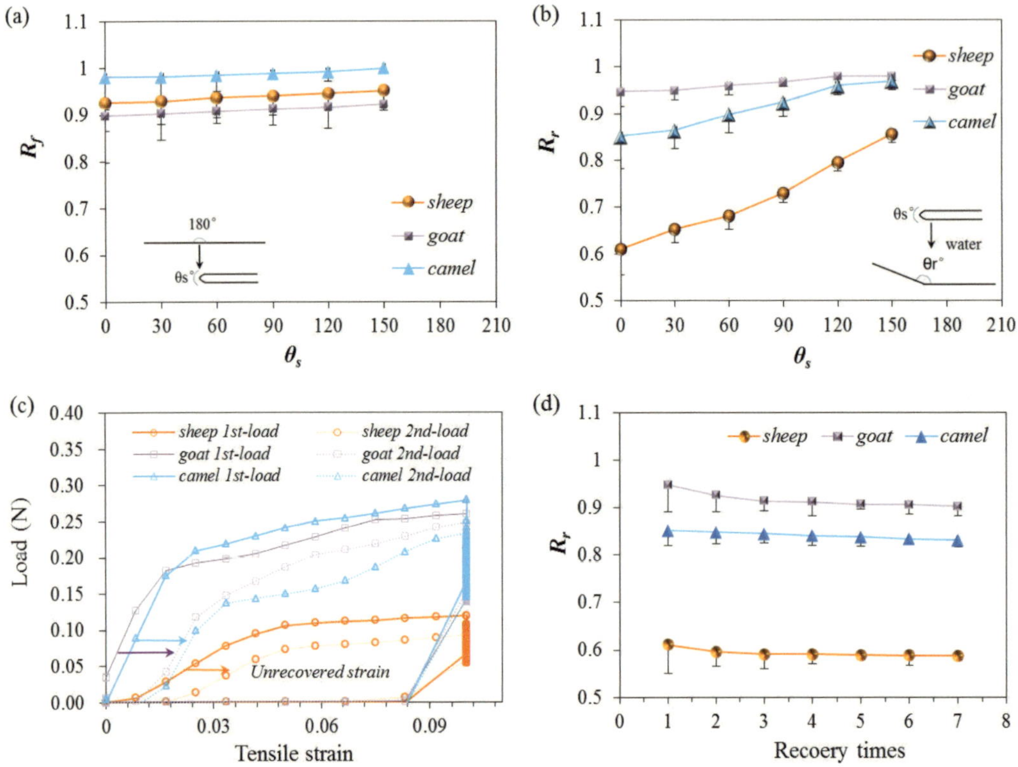

Figure 3. SME quantification of three keratin fibers responsive to water: (a) fixation ratio (R_f) values corresponding to a set of folded angles (θ_s) in temporary shape; (b) shape recovery ratio (R_r) from temporary shape to the hair innate shape (the inset illustrates the process of a camel hair from θ_s to θ_r in water); (c) two tensile cycles of three hairs responsive to water in shape recovery; and (d) R_r values of three animal hairs from completely deformed shape to the innate shape with the cyclic number of SME program [7].

Physically, a greater R_f value means a higher sensitive switch to be on and off, whereas a higher value of R_r implies a better SME of hair fiber. **Figure 3a** shows the measured R_f values of three animal hairs greater than 0.9, indicating the hairs with good penetration of aqueous molecules for hydration and high ability of removal of aqueous molecules for shape fixation that both relate to HBs in amorphous regions under original and dislocated states, respectively. An increased R_f value with an increase of set folded angles reflect that an enlarged set angle has less effect of folding process on hair molecule network, is found. In comparison with the other two hairs, the highest R_f value of camel hair may be due to the most normalized amounts of HBs in the cortex for the best shape fixation ability. The goat hair was measured to show the highest R_r value, indicating the hair with the best SME, as shown in **Figure 3b**. An increasing tendency for the relationship of R_r and θ_s is displayed for the measured keratin fibers. Among the parameters, R_r is determined by the entropic stress from the stable net-points and mutually connected network in the hair. The measured sheep hair shows the lowest R_r value which may be ascribed to the weakest network and least number and type of net-points for the lowest recover force. In **Figure 3a**, the dry sheep hair shows the thinnest cortex that may be the evidence for the reason of the lowest R_r value. Conversely, the goat and camel hairs manifest rapid recovery (less than 5 seconds faster than sheep hair) that indicates the larger amount or higher density of net-points/network (either crystalline phase or DBs) for stronger recovery stress.

On the other hand, the SME of keratin fibers can also be measured through cyclic tensile loads to keratin hairs by means of measuring the recovered strain using water dropwise to the fixed stretched hair. The mechanism of this method is to identify the elastic recovery, plastic recovery, and permanent unrecovered strains, in which the set tensile strain (0.10) is beyond the hair stretching yield point (around 0.05). As shown in **Figure 3c**, the elastic recovery takes place instantly when the elastic strain was released backward the original point and the unrecovered strain is thereafter found at the onset of the second tensile curve. In comparison to cyclic tensile curves, it is found that sheep hair also shows the most unrecovered strains after water dropwise to the hair and the goat hair still shows slightly less unrecovered strain than the camel hair, which is consistent with the triangle measurement results as shown in **Figure 3b**. Especially, at the strain of 0.10, the remarkably larger tensile loads for goat and camel hairs than the sheep hair support the assumed structure of stronger networks and more net-points of both keratin fibers. **Figure 3d** shows the corresponding R_r values of three keratin hairs for seven cycles of SME programs at $\theta_s = 0$, indicating that the shape recovery ability of keratin fiber stimulated by water is almost invariable (±5%) and the SME of keratin fiber is stable using water/moisture as stimuli.

3. Coupled thermal-water-induced shape memory behavior

Straightforward from **Figure 4**, a bundle of straight (Ori. dry) keratin fibers of camel hairs deforms into the helical structure (Def. heated) from the entanglement of raw straight hairs onto a circular steel bar (Def. dry) using a heating process. The deformed hair fibers in a free state maintained highly spiral residues (Fix. dry) that may be due to the heating on dislocated macromolecules, indicating the keratin hair with high shape fixation ability under thermal stimulus. However, the poor shape recovery ability was noted for the shape deformed hair using the same thermal stimulus [9]. In contrast, another water stimulus can recover the heat-deformed keratin hairs to their innate shape easily (*Rec. wet*), as shown in the recovered hair bundle in **Figure 4**. This means that the combination of factors of heat and water as coupled stimuli can give rise to good shape fixation ability and recovery ability; in another meaning, this can enrich the types of stimuli for keratin hair under processing and specific applications.

Here, **Figure 4** shows the SEM images of camel hair in microscale, in which the variation of hair cross section under different stimuli conditions reflects the coupled stimuli of SME of the hair. For the dry raw hair, it is featured with three components (**Figure 2a**). A large amount of macrofibrils show parallel alignment under the hair surface scales (cortex layer), which echo the hair hierarchical structural model. The observed macrofibrils show a key role in corresponding hair SME. The heating process can remove most aqueous molecules out of fibrils such that a decreased cortex volume is noted with more porous feature in morphology. In contrast, a wetting process to the deformed hair after heating can penetrate a huge amount of water into the macrofibrils, resulting in an obvious swelling phenomenon in lateral for the cortex and a decreased volume of hair central medulla. The reason was inferred to be the plasticization

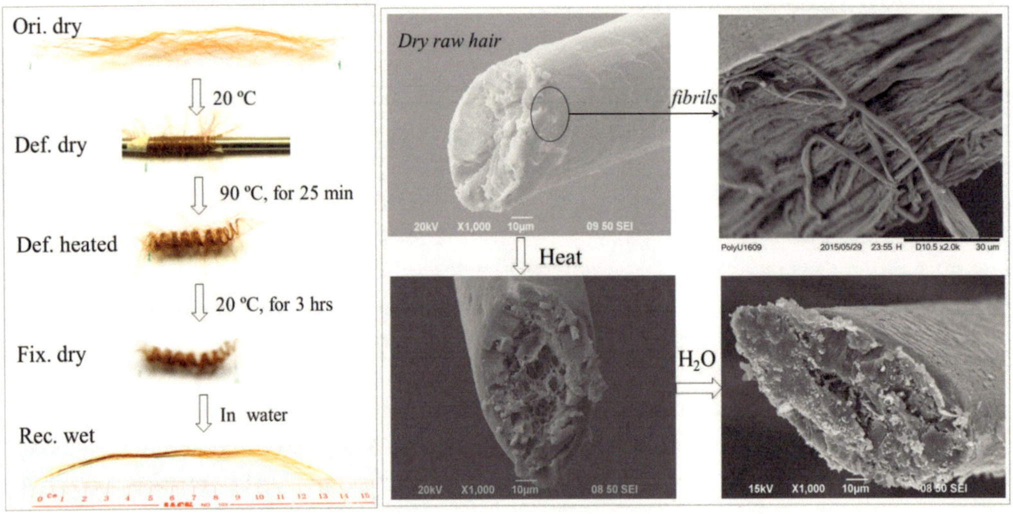

Figure 4. Demonstration of a bundle of camel hairs in coupled thermal-water-induced SME from innate shape (Ori. dry) to the heat processing on temporarily fixed shape (Def. heated and Fix. dry) and water-induced shape recovery (Rec. wet) that hydration effect of swelling is shown by SEM images of cross sections of a camel hair in dry and wet states.

of inter- and intra-fibrils, and the neighboring polypeptide chains were enlarged to their distance by means of disrupting HBs between imino and carbonyl (C=O, etc.) groups at adjacent branches. The water molecules enhance the mobility of adjacent macromolecules that seems like a key to switch on the temporary shape to recover.

Compared with water-sensitive shape memory behavior for hair shape fixation (**Figure 3a**), heat gives rise to a relative poorer ability in this respect. However, the measured R_f values (Eq. 1) of more than 0.75 indicate good shape fixation ability when responsive to heat, as shown in **Figure 5a**. The R_f value is increased as the increase of set folded angles (θ_s) implies a less effect of folding process on macromolecules network when the set angle is enlarged. The highest R_f (0.87) of hair under heating process may be ascribed to the less normalized amounts of regenerated HBs in the deformed camel hair cortex. The large error bar on each measured R_f value indicates the uneven hair shaft used in the folding SME characterization. The measured R_f values indicate that the deformed hairs under heating environment (90°C and 65% of RH) may undergo a transfer of aqueous molecules from the hair outside into the hair cortex that relates to the regeneration of HBs at the dislocated macromolecules, as shown in the schematic illustration of **Figure 5d**. Moreover, **Figure 5a** also displays the measured R_r values corresponding to R_f values. The R_r values of more than 0.87 indicate the excellent shape recovery due to the good hydration effect of deformed hair. In addition, the tendency of R_r and θ_s shows an increasing relationship, in which the R_r value is associated with the entropic stress from the stable netpoints and connected network. The lowest R_r of the hair ($\theta_s = 0$) may indicate the most setting effect on hair under high humidity and temperature, meaning the least number/type of netpoints for shape recovery. **Figure 5b** shows the relationship of measured R_f and R_r values ($\theta_s = 0$) after seven cycles of shape memory tests induced by coupled thermal and water, indicating the invariable shape fixation and recovery abilities of keratin fibers (decreased <10%).

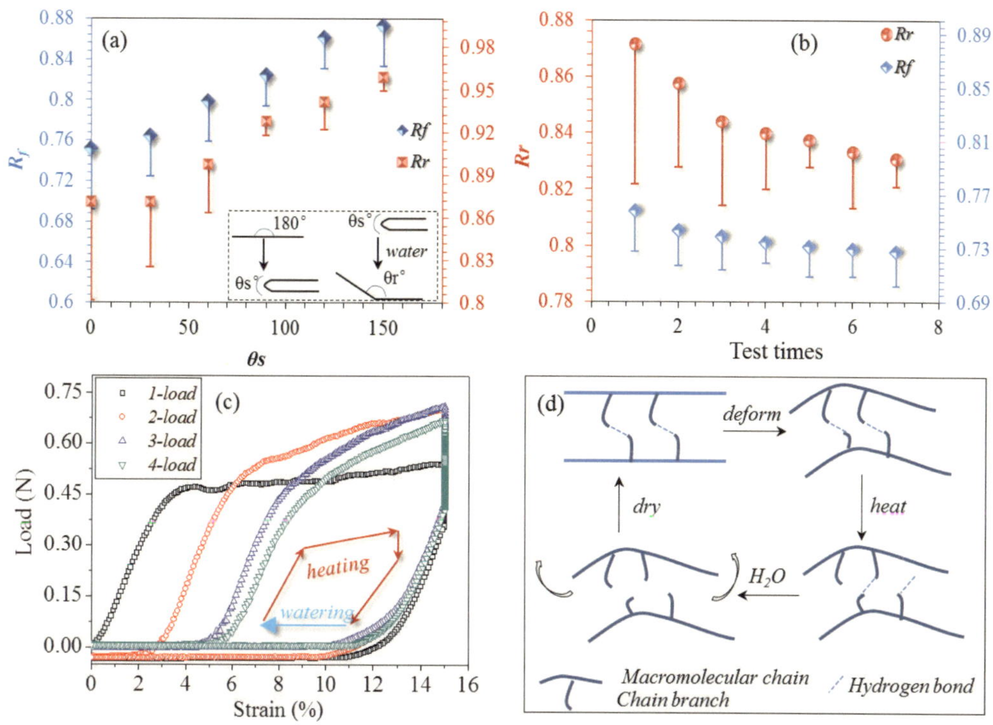

Figure 5. SME quantification of heat-induced shape fixation and water-stimulated recovery of hair fibers: (a) R_f in temporary shape and R_r from temporary to the innate shape corresponding to a set of θ_s, (b) variation of R_f and R_r values from $\theta_s = 0$ to the innate shape with the cyclic number of SME program, (c) cyclic tensile of an animal hair under four times of loading versus tensile strain with the coupled heat and water effects on SME of keratin fiber, and (d) schematic interpretation of coupled heat and water effects on the switch of keratin hair in macromolecular scale.

Cyclic tensile to keratin fiber is also employed here to measure the coupled stimuli SME. The 0.15 of set tensile strain is also beyond the yield point of the hair (~0.03). The shape memory strain can be identified from the unrecovered strain and the onset of the second tensile curve. Regarding the tensile curves in **Figure 5c**, the largest unrecovered strain should be the "four-load" curve which means that the hair has been experiencing four cyclic tensile programs. It is found that the unrecovered strain between two adjacent loadings is getting less with an increase of loading numbers. A calculated ~0.82 of R_r value is almost consistent with the measured results, as shown in the comparison in **Figure 5(a and b)**. Under the same strain of 0.15, it is found that the stretching forces of the "two-load" to "four-load" curve are remarkably greater than the "one-load" curve, supporting the viewpoint of realignment of macromolecule chains for stronger networks and net-points under conditions of tensile at 90°C.

4. Multi-responsive shape memory behavior

Figure 6a gives a normally used SME program for a keratin hair. In this section, four types of stimuli were used to investigate the multi-stimuli-responsive SME behavior of camel hair.

The related results were shown in **Figure 6b** with key images. Here, the water and redox agents as stimuli for the hair show remarkably SME, which the final recovered hair is similar with its original shape. For both factors, the aqueous molecules are believed to take an important role in SME of keratin hair [7]. In addition, as to the redox agent, the reductant ion, SO_3^{-2}, has broken down some DBs between keratin macromolecules after reducing reaction; thus, one type of net-point might disappear after shape fixation. The relatively weaker shape recovery from stimulus of redox confirms the viewpoint. Here, the stimulus factors, UV light and heat, both fix camel hairs with good temporary shape with high R_f values; however, the deformed shapes under the same environmental stimuli, i.e., poor R_r values, which can be comparably noted from the unrecovered residues of spiral circles after recovery processes are found difficult to be recovered. Theoretically, the camel hair samples were found in much drier state caused from stimuli factors of UV light and heat, where UV light breaks down the DBs and heat leads to less structural water [37], giving rise to the temporarily deformed hairs relatively difficult to be recovered in comparison with water and redox as stimuli under the recovery duration.

Duc to different growth environments and maturity, keratin hairs in fact manifest different shape memory abilities under different stimuli, depending on the hair internal contents, structure, and distribution of crystalline phase, HBs, DBs, etc. Quantitatively, according to Eqs. 1 and 2 for **Figure 7a–d**, the calculated R_f values, 0.84 and 0.875, indicate that the highest fixed strains can be achieved using water and redox as stimuli, respectively, which show higher values than the R_f values from UV (0.65), heat (0.73), and the original state (0.65).

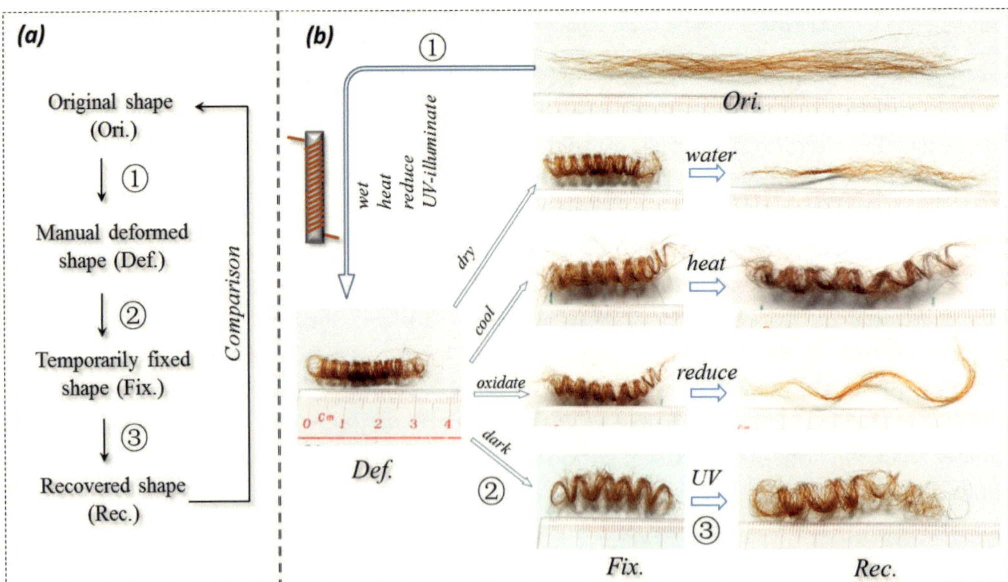

Figure 6. (a) SME investigation program and (b) observed SMEs of camel hairs under four stimuli (Ori. (original), Def. (deformed), Fix. (fixed), Rec. (recovered)) induced by water, heat (85°C), redox ($NaHSO_3$/H_2O_2 solutions), and UV light (254 nm), respectively [9].

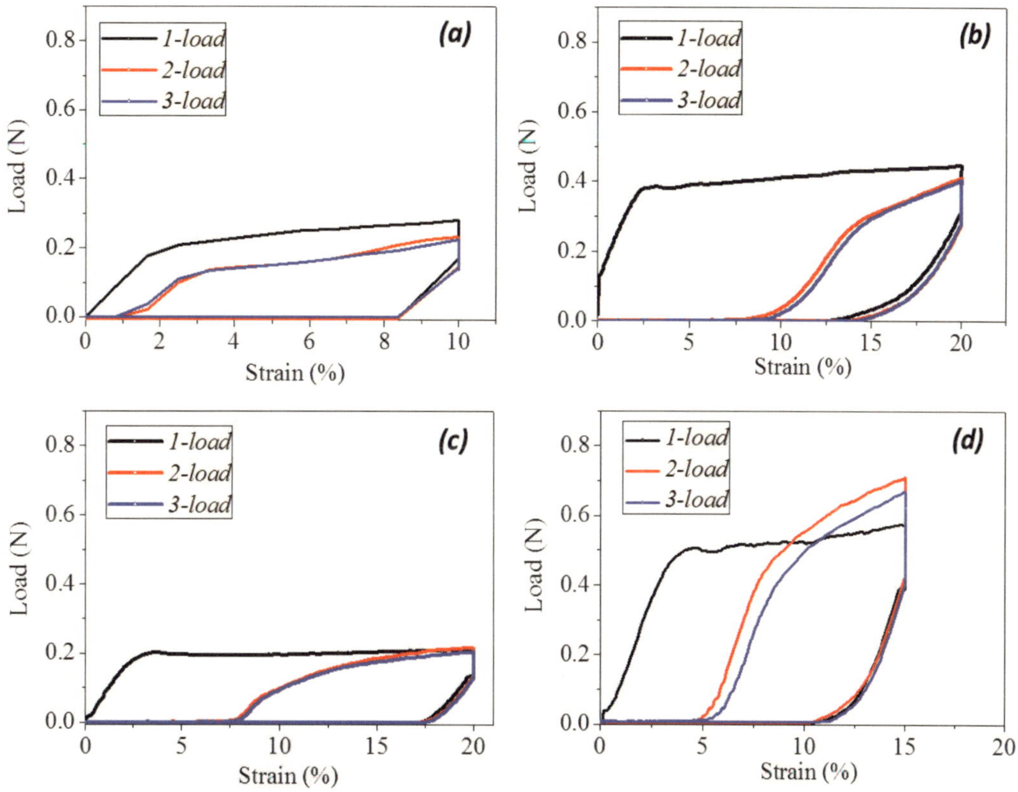

Figure 7. Experimental tests of cyclic tensile loads of camel hairs under (a) water stimulus, (b) stimulus of UV illumination (254 nm of light wave length), (c) stimulus of $NaHSO_3$-H_2O_2 redox solutions, and (d) heat (85°C) stimulus, respectively [9].

The reason may be the increased amount of HBs inside the hair from penetrated water molecules that act as switch for locking temporarily deformed shape. As to the R_r values, it was noted that the best shape recovery ability is from water (0.76), followed with stimuli of heat (0.547), redox (0.525), and UV radiation (0.225), respectively. Here, all shape fixation and recovery processes were under the same tensile speed for each stimulus. Water, thus, switches on the largest amount of HBs that causes the highest mobility of macromolecules and the most recovery strains from the released stress between net-points. To some extents, stimuli of heat, redox, and UV radiation can switch on the related reversible bonds in different degrees, leaving different recovered strains as shown in **Figure 7a–d**. The deformed hair can also recover slightly without constrain, and stimuli are ascribed to the relaxation of polypeptide chains. However, inevitably, the unrecovered strain due to plastic tensile cannot be completed in a short time. Partial DBs can be broken down when exposure of camel hair in UV radiation promotes the shape recovery as shown in the increased 3.5% of subsequent recovered strain based on the original hair. Similarly, the stimuli of heat and redox agent both can break down related HBs that give rise to more recovered strain for higher SME, indicating that HBs should be the key switch for SME of keratin animal hairs.

5. Identification of net-points and switch of keratin fibers in shape memory behaviors

Under UV radiation, reducing agent, heating, and watering conditions, the variation of internal components and structures of keratin fiber samples has been investigated from a number of characterization aspects. The DSC scanning shows two thermal absorption peaks for raw camel hairs, as shown in **Figure 8a**. The peak configuration displays that the first broad endothermic one appears at the temperature of 78.8°C with enthalpy of 270.1 J•g^{-1} that relates to the hair internal water removal from the hair [48]. The second DSC singlet appears at 230°C, while the inset figure showing the decreased 15.1 J•g^{-1} of enthalpy is ascribed to the denaturation of ordered α-helical keratin molecules [49]. This indicates that the crystalline phase is existed in animal hair, and its volume content is around 10–20% of the whole fiber, which would act as net-points in keratin hair's SME. **Figure 8b** shows five XRD patterns of camel hairs, in which the diffraction shoulders and peaks are noted for the original and arise by radiation (UV),

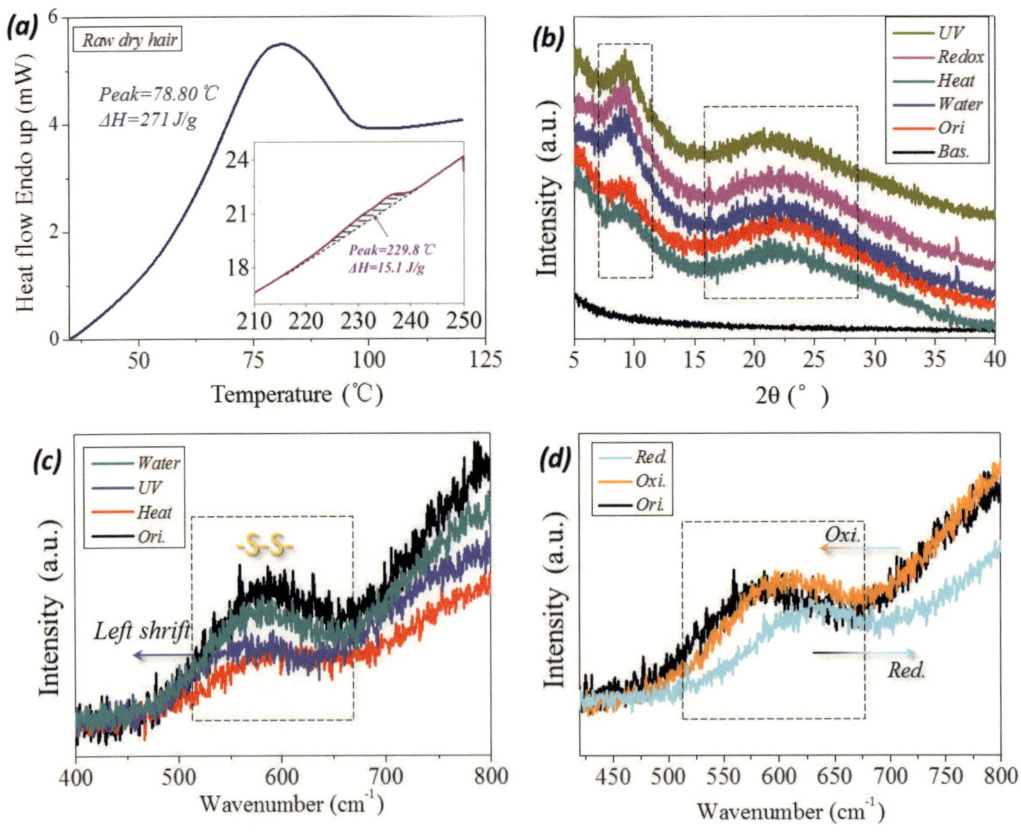

Figure 8. Identification of net-points and switches of keratin hair in SME: (a) DSC measurement of original camel hair (temperature ranging between 20 and 250°C); (b) XRD characterization of camel hairs in five states: Original, wet, heated, reduced, and UV radiated; (c) Raman spectra of camel hair samples after processing by water (20°C), UV radiation (254 nm of wave number), and heat (85°C), respectively; and (d) Raman spectra of camel hairs after reducing reaction (red. $NaHSO_3$, 1 M solution) and oxidizing reaction (oxi. H_2O_2 dilute solution) [9].

reduction (redox), heating (hea.), and hydration (wet) with sharp peaks, especially the peaks at the abscissa of $2\theta = 9°$(0.98 nm) and 21°(0.46 nm) that represent the α-helix and β-keratin crystalline phases, respectively. The almost identical intensities of both characteristic peaks for camel hairs at original (Ori.), heat, wet, redox, and UV indicate the unchangeable amount of crystalline phase during the five stimuli-induced shape memory behaviors.

In detail, some processes like UV radiation and water hydration in fact have a scarce effect on the crystalline phase as shown in the invariable crystal characteristic peaks. Therefore, the existed XRD peaks for α-helix of hair under multi-stimuli conditions indicate the net-point role of crystalline phase for related SME. In comparison to Raman spectra of camel hair samples under multi-stimuli ("wet," "UV," "heat," and redox ("red. and oxi.")) conditions, the original, wet, and heat curves can be viewed as nearly coincident in the Raman scanned regions (abscissa values of **Figure 8c** and **d**). Specifically, the DB mode is found between 500 and 580 cm^{-1} as a symmetrical broad characteristic peak that is associated with a few molecule conformations [10, 50] that are *g-g-g* (510 cm^{-1}), *g-g-t* (525 cm^{-1}), and *t-g-t* (540 cm^{-1}) (*g* and *t* denote *gauche* and *trans*) conformations, respectively. The processes of heating and hydration effects on hair DBs in amorphous area show ignorable with stable chemical cross-linking that may be net-points in related SMEs, as shown in **Figure 8c**. Nonetheless, the abscissa of the broad peak is noted to move forward to the right and then left side corresponding to the hair samples under stimuli of reducing agent and UV radiation subsequently. This indicates that the DBs of breakage and regeneration are different for both stimuli. However, the DBs have been proven to switch on after two stimuli on keratin fibers. The Raman spectra showing keratin hair with opposite motion of peak abscissa and intensity ratio indicate the symmetrical vibration of DBs in switch on (thiol groups) and off (DBs) states, suggesting that DBs can be switched in a few stimuli-induced SMEs.

For FTIR spectra, a broad absorption band at around 3400 cm^{-1} is noted corresponding to the hydroxyl group and free water molecules for wet hair samples [51], as shown in **Figure 9a** and **b**. Here, keratin hairs processed by water and reducing agent both show the deep absorption peak. Particularly, the key bands of keratin hairs such as Amide band I and Amide band II for both C═O stretching and N─H bending are found to shift to higher wave numbers from 1630 to 1633 cm^{-1} and from 1531 to 1533 cm^{-1}. These are shown with the HBs of band in dash frame in **Figure 9a** and **b** and the related measured shifting values in **Table 1**. This shifting motion of IR curves implies that the related HBs between intermolecules are re-formed from the dislocated HB residues and H_2O molecules during the hydration effect, being consistent with the reported results in [52]. Therefore, the penetrated H_2O molecules endow the keratin fiber with two existing waters, i.e., free motional and fixed bind molecules.

Regarding the SME program in detail, four IR curves for camel hair after processing by stimuli of heating (or UV radiation) for temporary shape fixation and by wetting (or reducing agent) for the final shape recovery are labeled in **Figure 9c**. An evident difference of the intensity ratio of characteristic peaks of N─H bending to C═O stretching vibrations for the camel hair under dry (innate or heated) and wet (in water or redox agents) states, which is consistent with the comparable curves in dash frame of **Figure 9a**, is noted. In addition, **Table 1** shows the calculated values of two characteristic peaks in wave number shifting and variation of intensity ratio along with the key SME steps under individual stimulus. In detail, the wave number of characteristic peak of C═O stretching is shifted by increasing

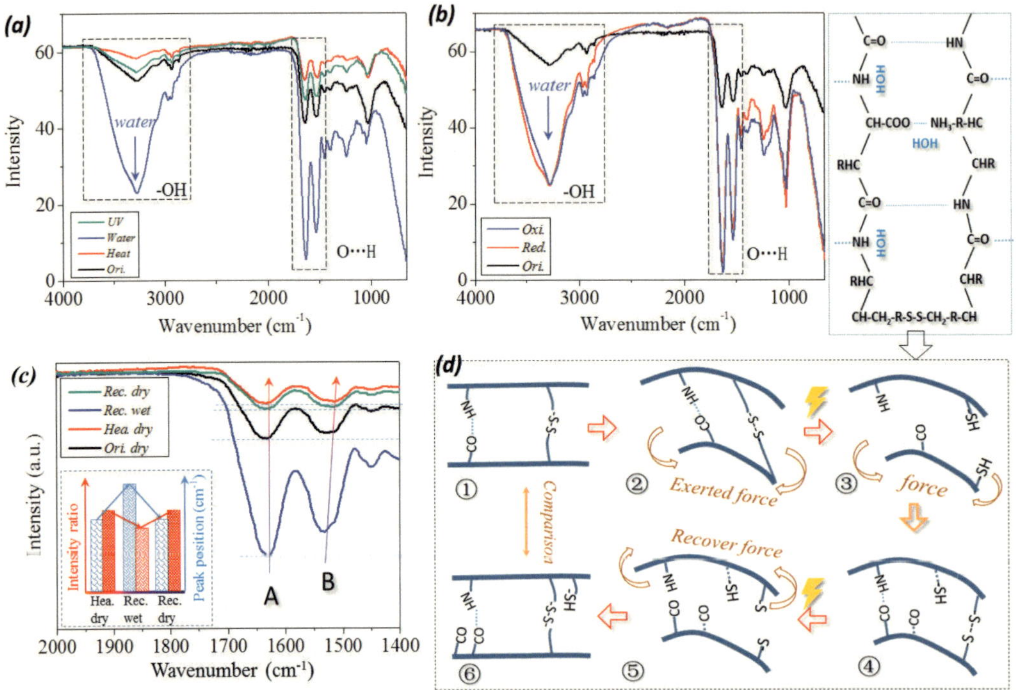

Figure 9. FTIR spectra of camel hair samples in the states of (a) Ori., water, heat (85°C), and UV radiation (254 nm of wave number); (b) red. and oxi. Conditions; (c) FTIR spectra for specific characteristic steps of SME program where the inset shows the IR characteristic peaks standing for HBs in open and close using shifting of peak abscissa and variation of peak intensity that also relates to the disruption and re-formation of HBs; and (d) simplification of keratin macromolecular chains and illustration of external stimuli on related switches in keratin macromolecule chains: ① original dry state, ② manually deformed dry state, ③ manually deformed state exposure on external stimuli, ④ temporarily fixed deformed state, ⑤ the fixed shape encounter with stimuli again, and ⑥ recovered state [9].

Hair states	IR peak position/wave number (cm^{-1})		Peak intensity ratio
Ori. dry/*Def. wet*	1630/*1633*	1531/*1543*	0.979/*0.854*
Ori. dry/*Def. heat*	1630/*1634*	1531/*1515*	0.979/*0.994*
Ori. dry/*Def. UV*	1630/*1629*	1531/*1532*	0.979/*0.988*
Ori. dry/*Def. red.*	1630/*1633*	1531/*1539*	0.979/*0.834*
Ori. dry/*Rec. oxi.*	1630/*1628*	1531/*1532*	0.979/*0.831*

Table 1. Characteristic IR peaks for camel hair under different processing conditions.

2–4 cm^{-1} from the dry hair to wet hair; conversely, the value is reduced by 2–4 cm^{-1} from wet hair to dry hair. In similarity of dry and wet conversion, the wave number shifting of characteristic peak of N—H bond move toward left and right by 8 and 16 cm^{-1}, respectively. This reversible conversion of the keratin hair under the original, heated, wet, and dry conditions, accordingly, indicates that the HBs in macromolecules experience a reversible destruction and reconstruction process [25]. However, the wave number shifting of the characteristic peaks of C═O and N—H vibrations has not been influenced by the stimulus of UV radiation, indicating that HB is not the switch for UV-induced SME of keratin fiber.

In detail to the variation of intensity ratio for the two characteristic peaks of keratin fiber, the value for camel hair has found an increase from dry state (Ori. dry, 97.9%) to heated (Def. hea., 99.40%), and wet (Rec. wet, 85.45% (in water) and 83.4 and 83.1% (in redox agents)) to dry (Rec. dry, 97.42%) reverses, respectively. **Figure 9d** interprets the conversion schematically based on the view of molecule motion corresponding to the six steps of four stimuli-induced SME program. The intensity ratios of two states, "Def. hea." and "Def. dry." (④), almost equal one, indicate that the intensities of peaks at 1510–1535 and 1620–1640 cm^{-1} equal each other and the amount of carbonyl and imino groups should be the same approximately. A heating or drying process would break the balance that the excess of free water molecules would be removed out of keratin fiber [53]. For example, a tiny amount of free water molecules were attracted onto N—H groups under normal moisture regain that may result in slightly decreased intensity ratio of characteristic peak for keratin hair at innate or deformed dry states from heated dry state (① and ⑥). The penetrated water molecules can break down HBs generated by carbonyl and imino groups (③ and ⑤) when the temporarily deformed keratin hairs were soaked in water or solution. In this case, each aqueous molecule was attached onto an imino group because of polar attraction between atoms N—H and H—O—H, giving rise to the decreased number of discrete N—H group for wave number of 1510–1535 cm^{-1} and correspondingly the evitable reduced intensity ratio for the two characteristic peaks. The temporarily locked shape from HBs was switched on by soaking process into free-constrained hair, and the shape recovered for the innate shape due to the released stress from net-points. The reversible conversion of drying and wetting of keratin hairs enable the same interaction of HBs among the hydrogen and oxygen atoms on amino, carbonyl, and hydroxyl in hairs, respectively [54]. Here, it should be noted that the interaction dynamics such as speed of responsive is not involved in the figure; the illustration only demonstrates the transformation process of each SME static step. In contrast, FTIR scanning cannot be used for characterizing the variation of DBs; thus, the SMEs induced by UV radiation and redox agents require another method to observe the variation, as referred to the DBs as switch under stimulus of redox in **Figure 8d**.

With respect to the tendency summary, as shown in the inset of **Figure 9c**, two characteristic IR peaks appearing at 1510–1535 and 1620–1640 cm^{-1} undergo the higher value of wave number shifting and reduced value of peak intensity ratio from the hair samples in the deformed heating/reducing or recovered wet to the final recovered dry state. The shifting and regular variation indicate that HBs in keratin fiber can lock its temporarily fixed shape by removal of internal water (heating or drying) and recover the innate shape wetting environment. This mechanism of HB implies it as the role of switch in water-, thermal-, and reducing agent-sensitive SME of keratin fibers.

After SME investigation of keratin fiber (taking camel hair as a typical keratin fiber), it can be concluded that keratin fiber is an intelligent biopolymer material with different shape memory functions that can be responsive to thermal, water, redox agent, and UV radiation in varying degrees. The inherent reason depends on the structural components of keratin fiber with different amounts of crystalline phase, HBs, and DBs among keratin macromolecules. To interpret the component role in SMEs, as shown in **Figure 10a** (original), a twin-net-switch

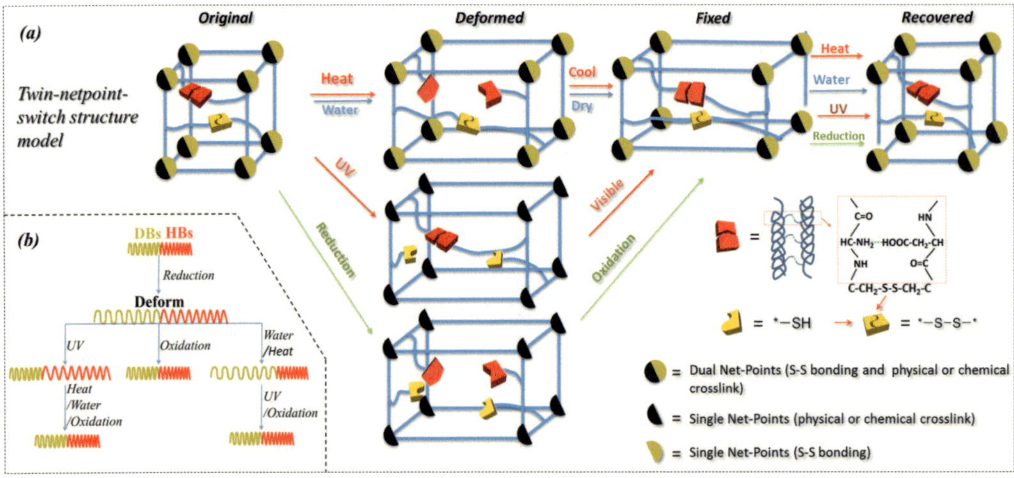

Figure 10. Multi-responsive SME mechanism: (a) a proposed structural model for different SMEs of keratin fiber; (b) the change of DBs and HBs under four types of stimuli where reducing agent can open both DBs and HBs; HBs can be regenerated using oxidant, water, and heat; and DBs can be re-formed using UV radiation and oxidant agent, respectively [9].

model is proposed that crystalline phase and DBs may be net-points and HBs and DBs may be switch, respectively. Therefore, two sets of net-points and switches for such smart biopolymer structure are, in simplicity, named as "twin-net-point-switch model." In the model, DBs act as two roles, depending on the environmental stimulus applied to the hair; for instance, DBs would be net-points for the hair in water, while they would be switch for hair in UV radiation or reducing agent. In addition, the keratin temporary shape can be programmed due to the opening of switch under a specific stimulus, and relative net-points can recover the innate shape. **Figure 10a** and **b** shows that HBs in amorphous region of the hair act as the single switch, whereas crystals and DBs both act as twin-net-points for stimuli of water and heat, DBs act as switch, and crystals work as single net-point for stimulus of UV radiation. Nevertheless, both bands are both switched on and off due to the water molecules and SO_3^{-2} ions for stimulus of reducing agent, indicating that they both are switch, while crystalline phase acts as single net-point without variation during SME program. With the aid of proposed model, a high possibility for developing an "all-in-one" intelligent SMP with multi-stimuli is believed to exist inspired from our rediscovery of keratin fibers.

6. Conclusions

In this chapter, keratin fiber was discussed for its SMEs responsive to various environmental stimuli, such as water, heat, coupled heat–water, UV radiation, and redox agent, respectively. The detailed mechanism was investigated using camel hair as a typical keratin fiber in molecular and structural networks. The keratin fiber with varying degrees of SME exposure on different stimuli for temporarily fixed shape recovering the innate shape was found. The calculated shape fixation ratio of more than 0.8 and recovery ratio of more than 0.5 indicate that the keratin fiber is a smart material stimulated by water and redox agent. The low shape fixation ratio (0.55) of keratin fiber under thermal stimulus and low recovery ratio (0.23) under UV radiation demonstrate that keratin fiber has less SMEs under such stimuli.

The cyclic tensile program of keratin fiber explains the SMEs well for each stimulus on the related switch on and off assuming net-point intact after stimuli. Some characterization approaches such as XRD and DSC of keratin fiber under five types of stimuli indicated that the crystals are invariable components during SME program. Raman spectra indicated the DBs in reversible breakage and regeneration when keratin fiber exposure on UV radiation and reducing agent, and invariation under stimuli of water and heat, indicates that DBs act as switch for stimuli of UV and redox agent and work as net-point for stimuli of water and heat. FTIR spectra of keratin fiber during SME program showed the related characteristic peaks in shifting wave number and variation of intensity ratio, indicating that HBs may act as switch for stimuli of water, heat, and redox agent. The last section proposed a structural model, namely, "twin-net-point-switch," for keratin fiber to interpret different SMEs of its exposure on different external stimuli, in which a twin-net-point/single-switch model is used for stimuli of water, heat and UV radiation, while a single-net-point/twin-switch model is used for stimuli of redox agent. This chapter for shape memory of keratin fiber is believed to have the ability to provide inspiration for manufacturing more remarkable synthetic SMPs responsive to multi-stimuli.

Acknowledgements

The authors gratefully acknowledge the financial support from the National Natural Science Foundation of China (Grant No. 51703083), the Natural Science Foundation of Jiangsu Province of China (Grant No. BK20160157), and the Hong Kong General Research Fund (RGC Project No. 15209815). In addition, the research support from Dr. Lu Jing, Mr. Wang Yifu, Miss Gui Xiaoting, and Dr. Wu You is specifically acknowledged in terms of experimental design and test, characterization and schematic drawing, etc.

Conflict of interest

Competing financial interests: The authors declare no competing financial interests.

Author details

Xueliang Xiao[1] and Jin-lian Hu[2]*

*Address all correspondence to: jin-lian.hu@polyu.edu.hk

1 School of Textiles and Clothing, Jiangnan University, Wuxi, P.R. China

2 Institute of Textiles and Clothing, The Hong Kong Polytechnic University, Hung Hom, Hong Kong

References

[1] Hu JL, Chen SJ. A review of actively moving polymers in textile applications. Journal of Materials Chemistry. 2010;**20**(17):3346-3355. DOI: 10.1039/b922872a

[2] Kumpfer JR, Rowan SJ. Thermo-, photo-, and chemo-responsive shape-memory properties from photo-cross-linked metallo-supramolecular polymers. Journal of the American Chemical Society. 2011;**133**(32):12866-12874. DOI: 10.1021/ja205332w

[3] Wang L, Yang XF, Chen HM, Yang G, Gong T, Li WB, Zhou SB. Multi-stimuli sensitive shape memory poly(vinyl alcohol)-graft-polyurethane. Polymer Chemistry. 2013;**4**:4461-4468. DOI: 10.1039/c3py00519d

[4] Tao Z, Peng K, Fan Y, Liu Y, Yang H. Multi-stimuli responsive supramolecular hydrogel based on Fe^{3+} and Diblock copolymer micelles Complexation. Polymer Chemistry. 2016;**7**(7):1405-1412. DOI: 10.1039/C5PY01742D

[5] Chen J, Zhang S, Sun F, Li N, Cui K. Multi-stimuli responsive supramolecular polymers and their electrospun nanofibers. Polymer Chemistry. 2016;**7**(17):2947-2954. DOI: 10.1039/C6PY00445H

[6] Schattling P, Jochum FD, Theato P. Multi-stimuli responsive polymers-the all-in-one talents. Polymer Chemistry. 2014;**5**:25-36. DOI: 10.1039/C3PY00880K

[7] Xiao XL, Hu JL. Animal hairs as water-stimulated shape memory materials: Mechanism and structural networks in molecular assemblies. Scientific Reports. 2016;**7**:26393. DOI: 10.1038/srep26393

[8] Hu L, Zhu Y, Huang H, Lu J. Recent advances in shape–memory polymers: Structure, mechanism, functionality, modeling and applications. Progress in Polymer Science. 2012;**37**(12):1720-1763. DOI: 10.1016/j.progpolymsci.2012.06.001

[9] Xiao X, Hu J, Gui X, Lu J, Luo H. Is biopolymer hair a multiresponsive smart material? Polymer Chemistry. 2017;**8**:283-294. DOI: 10.1039/C6PY01283C

[10] Akhtar W, Edwards H G M. Fourier-transform Raman spectroscopy of mammalian and avian keratotic biopolymers. Spectrochimica Acta Part A Molecular & Biomolecular Spectroscopy. 1997; **53A**(1): 81-90. DOI: 10.1016/S1386-1425(97)83011-9

[11] Harris M, Mizell LR, Fourt L. Elasticity of wool—Relation to chemical structure. Industrial Engineering Chemistry. 1942;**34**:833-838. DOI: 10.1021/ie50391a014

[12] Lindley H. The elastic properties of chemically modified wools and their implications for theories of wool structure. Textile Research Journal. 1957;**27**(9):690-695. DOI: 10.1177/004051755702700902

[13] Keten S, Chou CC, van Duin AC, Buehler MJ. Tunable nanomechanics of protein disulfide bonds in redox microenvironments. Journal of the Mechanical Behavior of Biomedical Materials. 2012;**5**(1):32. DOI: 10.1016/j.jmbbm.2011.08.017

[14] Chapman BM. Observations on the mechanical behaviour of lincoln-wool fibres supercontracted in lithium bromide solution. Journal of the Textile Institute Proceedings & Abstracts. 1970;**61**(9):448-457. DOI: 10.1080/00405007008630008

[15] Astbury WT, Woods HJ. X-ray studies of the structure of hair, wool, and related fibres. II. The molecular structure and elastic properties of hair keratin. Philosophical Transactions of the Royal Society of London. 1934;**232**(788):333-394. DOI: 10.1098/rsta.1934. 0010

[16] Samways C, Hastings GW. Radiation grafting of polysulphides to wool. Nature. 1970;**225**(5233):634-635. DOI: 10.1038/225634a0

[17] Lotay SS, Speakman PT. 3-Chain merokeratin from wool may be a fragment of microfibril component macromolecule. Nature. 1977;**265**(5591):74-276. DOI: 10.1038/265274a0

[18] Wu B, Yi Y, Xu T, Lei H. Characterization and functions of nano-wool crystallites. International Conference on Electronic & Mechanical Engineering and Information Technology. 2012;**62**(5):474-476. DOI: 10.2991/emeit.2012.311

[19] D'Hollander S, Assche GV, Mele BV. Novel synthetic strategy toward shape memory polyurethanes with a well-defined switching temperature. Polymer. 2009;**50**(19):4447-4454. DOI: 10.1016/j.polymer.2009.07.021

[20] Zhang TH, Wen ZB, Hui Y, Yang M, Yang KK, Zhou Q, Wang YZ. Facile fabrication of well-defined poly(p-dioxanone) dynamic network from metallosupramolecular interaction to access excellent shape-memory effect. Polymer Chemistry. 2015;**6**(22):4177-4184. DOI: 10.1039/C5PY00507H

[21] Ratna D, Karger-Kocsis J. Recent advances in shape memory polymers and composites: A review. Journal of Materials Science. 2008;**43**(1):254-269. DOI: 10.1007/s10853-007-2176-7

[22] Luo H, Liu Y, Yu Z, Zhang S, Li B. Novel biodegradable shape memory material based on partial inclusion complex formation between alpha-cyclodextrin and poly(epsilon-caprolactone). Biomacromolecules. 2008;**9**(10):2573. DOI: 10.1021/bm8004726

[23] Huang WM, Yang B, An L, Li C, Chan YS. Water-driven programmable polyurethane shape memory polymer: Demonstration and mechanism. Applied Physics Letters 2005;**86**(11):114105-114105-3. DOI: 10.1063/1.1880448

[24] Wortmann FJ, Rigby BJ, Phillips DG. Glass transition temperature of wool as a function of regain. Textile Research Journal. 1984;**54**(1):6-8. DOI: 10.1177/004051758405400102

[25] Eaves JD, Loparo JJ, Fecko CJ, Roberts ST, Tokmakoff A. Hydrogen bonds in liquid water are broken only fleetingly. Proceedings of the National Academy of Sciences of the United States of America. 2005;**102**(37):13019. DOI: 10.1073/pnas.0505125102

[26] Fan K, Huang WM, Wang CC, Ding Z, Zhao Y. Water-responsive shape memory hybrid: Design concept and demonstration. Express Polymer Letters. 2011;**5**(5):409-416. DOI: 10.3144/expresspolymlett.2011.40

[27] Jung YC, So HH, Cho JW. Water-responsive shape memory polyurethane block copolymer modified with polyhedral oligomeric silsesquioxane. Journal of Macromolecular Science Part B. 2006;**45**(4):453-461. DOI: 10.1080/00222340600767513

[28] Liu G, Guan C, Xia H, Guo F, Ding X. Novel shape-memory polymer based on hydrogen bonding. Macromolecular Rapid Communications. 2010;**27**(14):1100-1104. DOI: 10.1002/marc.200600189

[29] Chen H, Liu Y, Gong T, Wang L, Zhao K. Use of intermolecular hydrogen bonding to synthesize triple-shape memory supermolecular composites. Rsc Advances. 2013;**3**(19): 7048-7056. DOI: 10.1039/C3RA00091E

[30] Ware T, Hearon K, Lonnecker A, Wooley KL, Maitland DJ. Triple-shape memory polymers based on self-complementary hydrogen bonding. Macromolecules. 2012;**45**(2):1062. DOI: 10.1021/ma202098s

[31] Yao M. Study on the countermeasures of China-EU textile trade frictions. In: Advances in Education and Management. Berlin Heidelberg: Springer; 2011. pp. 148-153. DOI: 10.1007/978-3-642-23062-2_22

[32] Wedemeyer WJ, Welker E, Mahesh Narayan AM, Scheraga HA. Disulfide bonds and protein folding. Biochemistry. 2000;**39**(15):4207. DOI: 10.1021/bi9929220o

[33] Zheng M, Åslund F, Storz G. Activation of the OxyR transcription factor by reversible disulfide bond formation. Science. 1998;**279**(5357):1718-1721. DOI: 10.1126/science.279.5357.1718

[34] Yan Y, Wang Y, Heath JK. Cellular Association and Cargo Release of Redox-Responsive Polymer Capsules Mediated by Exofacial Thiols. 2011. DOI: 10.1002/adma.201101609

[35] Kakizawa YA, Harada A, Kataoka K. Environment-sensitive stabilization of core–shell structured polyion complex micelle by reversible cross-linking of the core through disulfide bond. Journal of the American Chemical Society. 2010;**121**(48):11247-11248. DOI: 10.1021/ja993057y

[36] Dailing EA, Nair DP, Setterberg WK, Kyburz KA, Yang C. Combined, independent small molecule release and shape memory via nanogel-coated thiourethane polymer networks. Polymer Chemistry. 2016;**7**(4):816. DOI: 10.1039/C5PY01464F

[37] Otsuka H, Nagano S, Kobashi Y, Maeda T, Takahara A. A dynamic covalent polymer driven by disulfide metathesis under photoirradiation. Chemical Communications. 2010;**46**(7):1150-1152. DOI: 10.1039/b916128g

[38] Canadell J, Goossens H, Klumperman B. Self-healing materials based on disulfide links. Macromolecules. 2011;**44**(8):2536-2541. DOI: 10.1021/ma2001492

[39] Fairbanks BD, Singh SP, Bowman CN, Christopher N, Anseth KS. Photodegradable, photoadaptable hydrogels via radical-mediated disulfide fragmentation reaction. Macromolecules. 2011;**44**(8):2444-2450. DOI: 10.1021/ma200202w

[40] Amamoto Y, Otsuka H, Takahara A, Matyjaszewski K. Self-healing of covalently cross-linked polymers by reshuffling thiuram disulfide moieties in air under visible light. Advanced Materials. 2012;**24**(29):3975-3980. DOI: 10.1002/adma.201201928

[41] Jakobs RTM, Ma S, Sijbesma RP. Mechanocatalytic polymerization and cross-linking in a polymeric matrix. Acs Macro Letters. 2013;**2**(7):613-616. DOI: 10.1021/mz400201c

[42] Dan A, Teramoto Y, Nishio Y. SH-containing cellulose acetate derivatives: Preparation and characterization as a shape memory-recovery material. Biomacromolecules. 2007;**8**(12): 3749-3757. DOI: 10.1021/bm7006828

[43] Chang R, Huang Y, Shan G. Alternating poly(lactic acid)/poly(ethylene-co-butylene) supramolecular multiblock copolymers with tunable shape memory and self-healing property. Polymer Chemistry. 2015;**6**(32):5899-5910. DOI: 10.1039/C5PY00742A

[44] Ponyrko S, Donato RK, Matejka L. Tailored high performance shape memory epoxy-silica nanocomposites. Structure design. Polymer Chemistry. 2016;**7**(3):560-572. DOI: 10.1039/C6PY90066F

[45] Feughelman M. A two-phase structure for keratin fibers. Textile Research Journal. 1959; **29**:223-228. DOI: 10.1177/004051755902900305

[46] Feughelman M. Natural protein fibers. Journal of Applied Polymer Science. 2001;**83**(3): 489-507. DOI: 10.1016/B0-08-043152-6/01037-8

[47] Mendez J, Annamalai PK, Eichhorn SJ, Rusli R, Rowan SJ. Bioinspired mechanically adaptive polymer nanocomposites with water-activated shape-memory effect. Macromolecules. 2015;**44**(17):6827-6835. DOI: 10.1021/ma201502k

[48] Tonetti C, Varesano A, Vineis C, Mazzuchetti G. Differential scanning calorimetry for the identification of animal hair fibres. Journal of Thermal Analysis & Calorimetry. 2015;**119**(2):1445-1451. DOI: 10.1007/s10973-014-4247-8

[49] Wortmann FJ, Deutz H. Characterizing keratins using high-pressure differential scanning calorimetry (HPDSC). Journal of Applied Polymer Science. 1993;**48**(1):137-150. DOI: 10.1002/app.1993.070480114

[50] Richard-Lacroix M, Pellerin C. Accurate new method for molecular orientation quantification using polarized Raman spectroscopy. Macromolecules. 2013;**46**(14):5561-5569. DOI: 10.1021/ma400955u

[51] Yao J, Liu Y, Yang S, Liu J. Characterization of secondary structure transformation of stretched and slenderized wool fibers with FTIR spectra. Journal of Engineered Fibers and Fabrics. 2008;**3**(2):1-10. WOS:000264531600003

[52] Ishida Y, Chabanne L, Antonietti M, Shalom M. Morphology control and photocatalysis enhancement by the one-pot synthesis of carbon nitride from preorganized hydrogen-bonded supramolecular precursors. Langmuir The ACS Journal of Surface & Colloids. 2014;**30**:447-451. DOI: 10.1021/la404101h

[53] Li G, Meng H, Hu J. Healable thermoset polymer composite embedded with stimuli-responsive fibres. Journal of the Royal Society Interface. 2012;**9**(77):3279-3287. DOI: 10.1098/rsif.2012. 0409

[54] Chen S, Hu J, Zhuo H. Properties and mechanism of two-way shape memory polyurethane composites. Composite Scienec and Technology. 2010;**70**(10):1437-1443. DOI: 10.1016/j.compscitech.2010.01.017

Chapter 7

Human Hair as a Testing Substrate in the Era of Precision Medicine: Potential Role of 'Omics-Based Approaches

Henry Ademola Adeola, Jennifer Cathryn Van Wyk,
Afolake Temitope Arowolo and
Nonhlanhla Patience Khumalo

Additional information is available at the end of the chapter

http://dx.doi.org/10.5772/intechopen.77215

Abstract

Minimally and noninvasive investigation of pathology and treatment monitoring is highly attractive in medicine. The use of human hair samples as a non-invasive testing substrate is potentially poised to improve diagnostic and forensic medicine. Hair has the unique ability to capture long-term information about health and disease in an individual as compared to urine and blood. Testing long hair offers a potential means of long-term monitoring of drug compliance, drug abuse, chronic alcohol abuse, and diagnostic biomarker discovery. Even though human hair is mostly composed of keratin and keratin-associated proteins, very little literature has been published on human hair proteomics. Emerging high throughput omics based techniques such as proteomics are increasingly improving our depth of knowledge about the diagnosis, prognosis and prediction of diseases globally. Although many aspects of the use of these novel molecular aids to improve disease diagnosis and patient management remains elusive; it is evident that these techniques have improved precision medicine tremendously. This chapter aims to discuss current plausible application of human hair omics-based approaches to the field of pathology, diagnostics and precision/individualized medicine.

Keywords: human hair, proteomics, precision medicine, hair testing, diagnosis, forensics

1. Brief history and background of human hair testing

The history of hair analysis dates back to as far as the nineteenth century. Precisely in 1858, Hoppe published a report on the discovery of arsenic in the hair of a human corpse exhumed after 11 years [1, 2]. Almost a decade later, amphetamine was discovered in the fur of a guinea pig by Goldblum et al., in 1954 [3]. In the late 1970s, Baumgartner et al., developed what is now known as a radioimmunoassay (RIA) kit for the detection of opiates in hair [4]. Thus, laying the groundwork for the first contemporary use of hair in drug testing. This method which was first introduced to Germany by Arnold in 1980 generated a lot of controversies [5]. However, subsequent work by Klug [6] that same year provided a basis for the use of hair in forensic toxicology by confirming the RIA method with thin-layer chromatography and fluorescence detection. Also, the inception of gas chromatography with mass spectroscopy (i.e. GC/MS) 6 years later, led to improved detection, sensitivity and specificity at low cost. Therefore, increasing the number of newly discovered compounds [1]. Over the next three decades, advances in chromatographic and spectrometric techniques coupled with new methods of sample preparation and wash procedures have enhanced the detection limits from ng/mg range to pg/mg [7]. Microscopic hair analysis, as another tool for hair testing, only began to gain a rapid recognition in the early twentieth century. For instance, the books published by Glaister in 1931 on the study of hairs and wools on mammals [8], and that of Hick on the microscopy of hairs in 1971 [9] both became a widely used resource for the forensic scientist in the use of hair as evidence in a crime scene. Preliminary findings from hair testing led to the American scandal of false imprisonment from evidence based on hair microscopy [10], which lead to severe scepticism on the validity of hair testing for analytical or forensic purposes.

Nonetheless, the scientific use of hair analyses and testing as a tool for forensic toxicology and medicolegal purposes, in the era of precision medicine, has proved valuable. More so, the recent need for a non-invasive method/testing substrates for early diagnosis and profiling of diseases based on their biology and molecular phenotypes [11], has made hair testing an attractive option. Also, application of non-invasive strategies improves patient compliance. Considering that the human hair is mostly composed of proteins; it is an attractive proteomic substrate for diagnostic and forensic pathology. Not least, the presence of other biomolecules (such as lipids) in hair indicates that an integrated multi-omics approach would significantly improve the field of investigative medicine [12].

It has been hypothesized that irrespective of ethnicity or hair form, there is biochemical and growth rate similarities in all hair types [13]. However, emerging evidence seems to suggest that differences exist in hair structure and growth rates depending on ethnic profile or grooming [14]. Human hair distribution and growth is androgen-dependent resulting in differences between males and females as well as regional variation within the same individual [15].

The human hair follicle contains stem cells in their bulge region and in the dermal papilla [16]. These stem cells are responsible for the significant self-renewal capabilities of the human hair follicle and the hair cycle. There are remarkable disparities in molecular structure and biology of human hair based on individual differences, gender, age, various hair grooming habits; chemotherapy/radiation exposure; cosmetics as well as diseases [17].

Omics-based techniques employ a holistic interrogation of the full complement of a biomolecule in a systems biology oriented manner. Although, many omics-based methods are yet to be employed routinely in hair testing and biomarker identification; there is reasonable hope that they will provide much needed insight to the biology and pathology of several medical conditions. This review intends to evaluate the current hair testing applications and potential benefits of hair omics-based approaches in diagnostic medicine.

http://dx.doi.org/10.5772/intechopen.77215

2. Basic structure and biochemistry of the human hair

Human hair is a long, thin cylinder of keratinized cells that grows from large cavities or sacs called follicles. Hair is usually composed of three distinct regions: an inner cortex, an external cuticle and occasionally a central inconsistent medulla or core running along a central axis of thicker terminal hairs [18]. The cortex is primarily responsible for the mechano-physical properties of the hair fiber and is composed of long filaments called microfibrils which contain organized α-helical rods of keratin, embedded in an amorphous matrix [19] (**Figure 1**). Human hair is composed of 65–95% proteins, 15–35% water and 1–9% lipids, 0.1–5% pigments

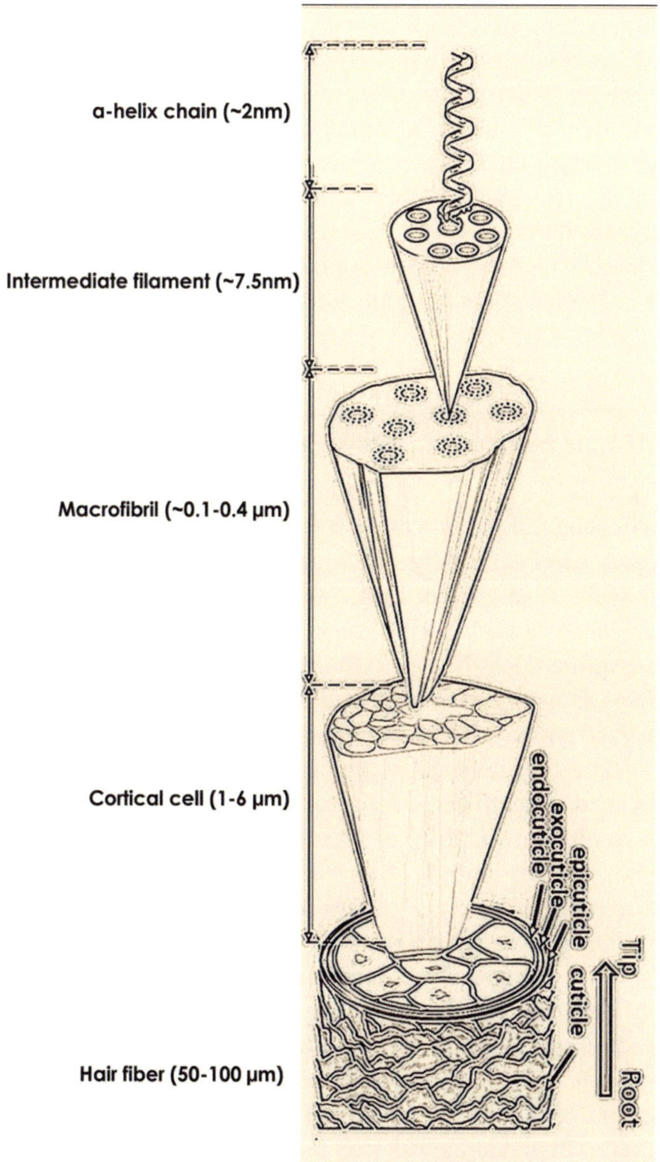

Figure 1. Complex structure of the human hair fiber (picture adapted with kind permission [128]).

(melanin), small amounts of trace elements, and polysaccharides [18]. Older studies suggest there is no significant difference in the amino acid composition of hairs from different racial group [20]; however, there is evidence that there are significant differences in the lipid composition of hair [21]. The hair fiber is rich in lipids either derived from sebum or secreted from the apocrine gland which consists of free fatty acids, mono-di-and triglycerides, wax esters, hydrocarbons, and alcohols [22].

On average, human scalp hair is reported to grow approximate 1 cm/month. However, recent data reports lower growth rates for hair originating from people of African ancestry. The human hair growth cycle consists of four phases called anagen (growth), catagen (transition), telogen (rest) and exogen (shedding phase) [23, 24]. During the anagen phase, the capillary blood supply around the follicle provides nutrients and delivers any extraneous materials that may be present in the blood stream such as drugs or toxins [10]. Although the mechanism of the incorporation of drugs or metabolites into hair is not clear, three pathways have been proposed and are generally accepted by scientists including: active or passive diffusion of drugs into the bloodstream feeding the dermal papilla, diffusion from sweat or other excretions the growing or mature fiber is exposed to and, external diffusion from vapors or powders into the mature hair fiber [25]. However, the relative importance of each pathway has not been elucidated and it is probable that the incorporation of drugs into hair may involve a complex series of events which may conceivably vary between different individuals.

3. Reliability of hair testing in precision medicine

In comparison with blood sample collection, hair collection is easy and non-invasive. It does not require any specialized equipment or storage facilities. It permits long-term storage at room temperature without any risk of contaminating the environment. There is reasonable evidence that using hair as a testing substrate in medicine is of tremendous benefit and that there are many unexplored possibilities to the use of hair testing in the evidence-based era of precision medicine. However, it is important to comment about the reliability of hair testing as this has far reaching ramifications in various scenarios. In a study that assessed the reliability of inter-laboratory and intra-laboratory variations in hair minerals using three different laboratories, a consistent numerical result was generated from all three laboratories [26]. However, the results were interpreted differently because each laboratory used different normal reference ranges. This indicates that variation in reported concentrations of specific analytes in hair would result in different interpretation of the patient's health status in each instance. Therefore, effective ancillary use of hair in diagnostic medicine requires standardization of normal reference ranges. Over- or under-reporting may be another factor that might affect the reliability of the use of hair as an alternative biological testing matrix. For instance, Vignali et al., found discrepancies between self-reported use and biological testing of illicit drugs use among inmates [27]. The authors explained that theses discrepancies might be due to false declaration of drug use in hope of qualifying for entry into rehabilitation programs. Also, illicit drug users with moderate risk may use levels below the limit of detection because

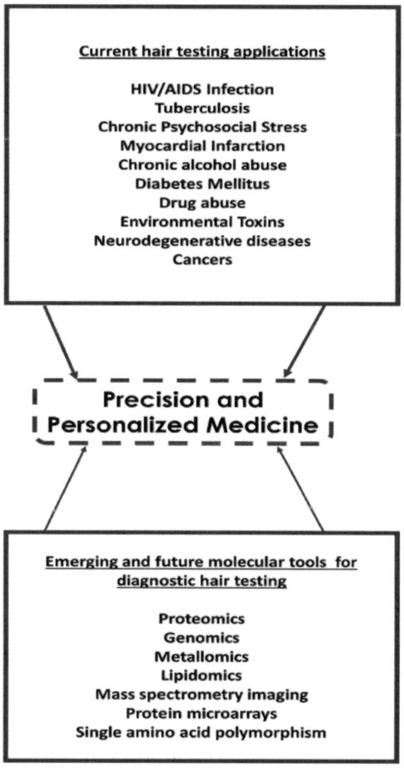

Figure 2. List of current application of hair testing in personalized medicine and future diagnostic potential of omics-based approaches for precision medicine.

there is inadequate empirical evidence of standardized minimum detectable levels of specific substances of abuse; hence, a negative test does not connote "no exposure" [28]. On the other hand, hair testing has become so reliable that it was considered better than urine, blood or breath testing, for demonstrating long-term illicit drug abstinence prior to solid organ transplant [29]. Hair testing may present a complementary test in scenarios where the veracity of conventional test results is in doubt. A summary of current and future application of hair testing in precision medicine is represented in **Figure 2**.

4. Current hair testing applications in medicine

The human hair is capable of long term incorporation of various exo- and endogenous compounds as compared with blood plasma or urine [30]. Changes in the physicochemical properties of hair as well as genetic, environmental and hormonal factors may result in various hair disorders [31]. Hence, a good understanding of the biology and the potential use of hair as a testing substrate in medicine is a very attractive option. Hair testing approaches have been applied for drug and biomarker monitoring in forensic and toxicological screening purposes as described below:

HIV/AIDS infection: Inadequate exposure to medication results in antiretroviral therapy failure, but unfortunately there are no robust methods for long-term monitoring of drug compliance in HIV/AIDS patients. Hair-based monitoring of protease inhibitors (like atazanavir and lopinavir) concentration have been reliably shown to be independently and strongly associated with treatment response [32]. Another study carried out by Hickey et al., demonstrated a relationship between nevirapine concentration and virological outcomes, although high levels of nevirapine found in women and older adults in their cohort require further research [33]. However, non-significant negative association between nevirapine and drug compliance has also been reported in a HIV-infected pediatric cohort [34].

Tuberculosis: Blood plasma monitoring of tuberculosis (TB) can be fraught with a lot of inconsistencies due to the relatively shorter half-life of the anti-TB drug in plasma as compared with hair. For example, low plasma levels of rifampicin, isoniazid, and pyrazinamide have been demonstrated in many pulmonary TB patients, despite compliance with the direct observation of treatment (DOT) strategy [35]. Inter-patient pharmacokinetics has made a reproducible assessment of drug compliance difficult in TB patients. More recently, adherence to anti-TB drug regimen has been monitored effectively in latent and active TB patients using a liquid chromatography coupled tandem mass spectrometry (LC-MS/MS) analysis to identify isoniazid levels in human hair samples [36]. Also, isoniazid levels have been monitored in pediatric population of TB patients using small quantities of their hair samples [37]. A landmark non-invasive multi-drug quantitative assay has also been developed, for measurement of multiple second line anti-TB drugs (moxifloxacin, pyrazinamide, linezolid, and levofloxacin) in multi-drug resistant types of TB, using hair sample testing [36].

Chronic psychosocial stress: Chronic stress has a detrimental impact on human health if allowed to persist indefinitely. An important physiological response to stress is the stimulation of the hypothalamic-pituitary-adrenal axis and release of cortisol [38]. Due to diurnal variation in cortisol release, measurement of cortisol levels in body fluids such as urine, blood and saliva may be fraught with inconsistencies [38]. Hence, hair cortisol level measurement is now being used as a surrogate biomarker for chronic psychosocial stress [38, 39].

Myocardial infarction: Acute myocardial infarction (MI) can be precipitated by acute stress, albeit the role of chronic stress in acute MI is poorly understood. Human hair testing is emerging a reliable method to measure cortisol levels around episodes of acute MI. An enzyme immunoassay-based approach has been used to demonstrate higher cortisol levels in the most proximal 3 cm portion of the hair of acute MI patients as compared to patients without acute MI [40].

Chronic alcohol abuse: Beside its use to detect alcohol consumption and monitor abstinence from alcoholic beverages, ethyl glucuronide a stable metabolite of ethanol has the capability for long term monitoring of alcohol abuse using hair samples. Whilst other tests such as breathalyzer, urine test, and blood test are capable of short term records of ethyl glucuronide (at most 3–4 days), hair is capable keeping track of months of alcohol abuse. The efficacy of hair testing for ethyl glucuronide as a biomarker of chronic alcohol abuse is well established [41].

Diabetes mellitus: Diabetes mellitus (DM) is a leading cause of morbidity and mortality globally; with global estimate of people with DM projected at 300 million by the year 2025 [42].

Long term damage, failure, and dysfunction of various body organs can result from the chronic hyperglycemia that characterizes DM [43]. Westernization of diet as well as lifestyle has been suggested to be a contributory factor to the development of DM in Africa [44]. As far back as 1960s and 1970s, researchers have investigated in various cohorts, the correlation between the level of chromium in human hair and DM [45]. Later on, hair protein glycation has also been used for long-term DM treatment monitoring. For example, glycated products such as furosine has been correlate fairly with the yearly mean values for HbA1c and fasting plasma glucose in a study by Oimomi et al. [46]. Also, a spectrophotometric hair glycation index has also been shown to be reliable to differentiate between normoglycemic and hyperglycemic individuals [47]. More recently, determination of derivatized amino acids in human scalp hair using gas chromatography-coupled mass spectrometry, revealed differential molecular signatures between DM patients and controls in a Jordanian cohort [48]. Thus, analysis of human hair is a promising approach to long-term evaluation of DM and other diseases.

Drug abuse: Hair testing is a key area of growing interest in the fields of forensic sciences because it can be sampled without difficulty from human subjects. Also, hair possesses the ability to store chronological record of drug use and abuse. Human hair testing has led to accurate measurement of levels of various illicit drugs such as amphetamines, cocaine, cannabis, opiates, phencyclidine, barbiturates and methamphetamine have been measured using radioimmunoassay and gas chromatography coupled mass spectrometry [49].

Assessment of exposure to environmental toxins: Environmental exposure to various toxins may accumulate and lead to various health problems. Normally, exposure to xenobiotic agents induces post-translational modifications (PTMs) to proteins [50], they may induce heat shock proteins [51] and lead to the induction of phase-I and phase-II detoxification enzymes [52]. Currently, exposure to environmental toxins (such as heavy metals, insecticides, polycyclic aromatic hydrocarbons (PAHs), endocrine disruptor pesticides, etc.) can be evaluated by the use of hair testing [53–57]. Although promising, hair testing for environmental toxin may be fraught with inaccuracies due to factors such as the lack of analytic validation methods, low interlaboratory reliability and the poor delineation between endogenous toxicants and exogenous contaminants in hair [58].

Neurodegenerative diseases (ND): Hair testing is gradually finding a role in molecular assessment of NDs [59]. For example, differential PTMs of proteins are manifested in response to ND. Two NDs, Menke's kinky hair syndrome and Elejalde disease are characterized by hair changes that may be of great diagnostic value [60]. Both conditions result in irreversible brain damage and early death and hair abnormalities seen in these patients help to differentiate these conditions from other NDs [60]. Scalp hair trace element content has been used to diagnose multiple sclerosis [61]. Also, red hair phenotypes have been associated with greater risk for Parkinson's disease [62]. Further molecular studies are needed to explore noninvasive diagnostic in the field of neurodegenerative medicine.

Cancers: As in the case with ND, differential PTMs and expression of proteins are manifested in response to cancer. Also differential expression of various trace elements in hair have been found for different types of cancer, in a study carried out among Polish [63] and Indian [64] populations. X-ray diffraction of hair, used in combination with mammography has been reported to significantly improve the sensitivity and diagnostic accuracy for invasive Breast Cancer [65]. A meta-analysis demonstrated that hair zinc levels in female breast cancer patients were lower than those found in controls, albeit this difference could not be picked up in serum [66]. Cancer-associated hair phospholipids can also be potentially used for development of a reliable screening hair-based test for breast cancer [67].

5. Future perspectives

5.1. Experimental Omics-based molecular hair testing

Various omics-based approaches have allowed researchers to interrogate the full complement of molecules in complex biological systems. Up to 73 species of ceramide lipids have been identified using shotgun lipidomics analysis [68, 69]. Also, there is a positive correlation between hair arsenic, sodium and iodine levels and risk of cancer development in a metallomic study [70].

Proteomic techniques have also been successfully utilized in hair [71–75]. Mass spectrometry based approaches have led to the identification of peptide biomarkers of oxidative chemical damage in hair, as well as to differentiate structural damage caused by cosmetic treatment/ weathering hair diseases like monilethrix [76]. Also, mass spectrometry has been used to identify 343 proteins as natural constituents of the human hair [73]. Using liquid chromatography-coupled mass spectrometry-based shotgun proteomics, ethnic-based differential proteomics signature have been identified in hair keratin protein levels [75].

Also, mass spectrometry imaging (MSI) has become a major tool for tissue biomarker discovery [77]; as well as other applications such as tissue profiling in histopathology and drug distribution in forensic medicine [78]. Waki et al., utilized MSI technology to determine biomarkers for aging in the human hair cortex [79]. MSI has been shown to be able to monitor cocaine in a single strand of hair sample for forensic purposes [80]. The concentrations of cocaine varied along the shaft of the hair strand, which correspond to the chronological time the drug was ingested.

A protein microarray proteomics study identified autoimmunity related biomarkers for scalp lesions in alopecia areata [81]. This study identified eight potential antigen biomarkers specific for alopecia areata which were used for the design of a discriminant miniarray for alopecia areata and other scalp lesion [81]. Not least, single amino acid polymorphism (SAP) or non-synonymous single nucleotide polymorphisms (SNPs) are emerging as a potential approach for reliable human identification for forensic purposes [71, 82].

An important feature of all omics-based methods is their high dependence on data repositories, algorithms and bioinformatics workflows [83]. A "public data driven" approach for integration, mining, and reuse of data would benefit the field of omics immensely [84]. Although there are individual merits for each omics based field, an integrative multi-omics method may permit a more robust molecular analysis [85]. Despite the few available papers that have addressed the use of omics-based methods in the field of dermatology [86, 87], there still remains a dearth of papers that address the application of proteomics to human hair samples for non-invasive testing of various clinical conditions for precision medicine.

Hair metabolomics: The term metabolome refers to all the metabolites in a biological organism, with metabolites being the products of enzyme-catalyzed reactions that result from gene expression, protein synthesis and environmental stressors [88]. Metabolomics is a relatively new and dynamic field [89] and involves the study of "the complete set of metabolites/low-molecular-weight intermediates, which are context dependent, varying according to the physiological, developmental or pathological state of the cell, tissue, organ or organism" [90]. Depending on the length of the hair, endogenous compounds and physiological influences can be observed in hair over extended periods of time. Utilizing hair metabolomics in personalized medicine allows clinicians to deliver optimum care to patients [91]. Advances in this field enable the accurate determination of possible pharmaceutic toxicity, appropriate dosages and an individual's susceptibility to possible disease. In addition to physiological influences, co-metabolic factors are considered when investigating hair metabolomics [129]. This enables the examination of drug metabolism during the different stages of disease. Metabolomics is increasing used to understand and predict drug response in patients. For example, a gas chromatography-time of flight (GC-TOF) based platform was used to investigate the biochemical differences associated with the racial response to atenolol, an anti- hypertensive drug [92]. Other fields of study often are unable to account for various day-to-day factors present in individual lives. Combining hair metabolomics with other disciplines and rapidly advancing analytical methods enables a holistic investigation into metabolism and drug metabolism. An example of this is the identification of biomarkers for spontaneous preterm birth using hair metabolomics in conjunction with blood biomarkers [93].

Hair lipidomics: Lipidomics is a branch of metabolomics focused on the complete analysis of lipid species and their biological roles in health and disease. Lipids are analyzed using mass spectrometry (MS), chromatography and spectroscopy-based methods. Hair lipids are distributed throughout the hair fiber, they are hydrophobic, oily or greasy organic substances that are extractable from cells and tissues by nonpolar solvents, mainly chloroform and/ or ether [94]. Lipids are designated as exogenous and endogenous depending on whether they emanate from sebaceous glands or hair matrix cells, respectively. Endogenous lipids, biosynthesized in the hair matrix cells accounts for approximately 2.5% of the entire hair fiber include cholesterol, sulphate, sterols and ceramides while exogenous lipids accounts for below 1% which include squalene and sterols [94] and is vital for maintaining the internal water content in the hair fiber [95].The application of lipidomics hair research is still in its infancy. MALDI-MS Redox lipidomics has been used to investigate lipid damage in human hair [96]. Studies showing the importance of lipids in in diseases such as alopecia [97], breast cancer [98, 99], and diabetes is sure to promote research in hair lipidomics.

5.2. Potential ancillary molecular tools for hair testing

5.2.1. Molecular imaging

Molecular imaging (MI) can be defined simply as the characterization and measurement of biological processes at the tissue; cellular and molecular level using optical techniques. In contrast to the 'traditional' diagnostic imaging, MI investigates the molecular abnormalities underlying a disease instead of the mere visualization of the end effects of the molecular aberrations [100]. Although the use of imagining in hair analysis is not new, the recent advancements have shown that the use of microscopy goes beyond the comparison and identification of microscopic characteristics in hair. More so, the combination of microscopy with spectroscopy techniques promises to be a valuable tool for the molecular characterization of hair [101, 102]. Molecular imaging of hair in comparison to other biological materials is very useful particularly in diagnosis as it provides a useful testing or screening material with minimal invasion to the subjects. Examples of imaging tools used in the analysis of hair apart from mass spectroscopy imaging include the following:

Multiphoton fluorescence lifetime imaging (MFLIM): MFLIM microscopy is a method set up for the optical examination of biological samples [103, 104]. Due to the inherent sectioning capability of this technique, 3D images with subcellular resolution can be obtained as well as two-photon excited auto-fluorescence of endogenous fluorophores. In 2007, Ehlers et al., studied the bleaching effects, intrahair dye accumulation and other hair pigmentation properties of the human hair using a time-resolved MFLIM single photon counting and near-infrared femtosecond laser pulse excitation [105]. Two photon imaging have also been employed in the live imaging of stem cells and the study of progeny cells in hair follicle regeneration [106].

Synchrotron radiation X-ray fluorescence imaging (SRX): The use of synchrotron-radiation-excited X-ray fluorescence to analyze biological samples provides information about the presence of trace metals [107]. The minimal radiation damage of SRX imaging to specimens in comparison to other techniques involving the use of ion or electron probes makes this method a preferred choice to study the distribution of elements in pathological condition and pharmaceutical intervention [108]. The application of SRX imaging to study the dynamics of mercury in rat hair [107] and mapping of other trace elements in human using a muprobe have been reported [109].

Fourier transform infrared (FTIR) imaging: FTIR imaging has an advantage over other imaging techniques in that it is label-free, nondestructive and has a broad range of biomedical application [110]. Most especially, the use of the FTIR imaging in the attenuated total reflection (ATR) mode [111]. A key advantage of the use of ATR-FTIR imaging is that there is minimal sample preparation before analysis and the depth of penetration of infra-red (IR) light in the specimen is well controlled and is independent of sample thickness. Thus, allowing for a good spatial resolution of the imaged samples and even faster image acquisition (two to three orders of magnitude) when combined with array detectors such as focal plane arrays (FPA) or linear arrays [110]. Significant improvement in the signal to noise ratio have been recorded with the use of synchrotron as source infrared radiation. However, the use of this method is only suitable for very small specimen areas and will involve a mapping approach for the imaging of larger sample area. In addition, the spatial resolution obtainable with synchrotron radiation source in FTIR imaging is not desirable. However, the advent of Globar infrared source has improved spatial resolution and allow for a high detection and distribution of the chemical composition of biological samples [112]. Thus, the application of FTIR-ATR imaging has been used to obtain clear chemical images of the cross-section of the human hair and thus obtain information from the medulla without the problem of passive contamination of the adjoining cortex [113]. The different forms of FTIR and other examples of molecular imaging techniques mostly involving a combination of IR or non-IR spectroscopy techniques that have been used in hair analysis and diagnosis of disease with hair as substrate includes FTIR spectral micro-imaging [114, 115] infrared and Raman spectroscopy [116], Atomic force infrared spectroscopy (AFM-IR) [117, 118] and electron microscopy [119–121].

http://dx.doi.org/10.5772/intechopen.77215

5.2.2. Nanotechnology/Nanomedicine

Nanotechnology involves the control of matter at the atomic or molecular levels, i.e. at the nanoscale (1–1000 nm). Hence, nanotechnology can be defined as methods or techniques employed in the processing of materials at the atomic or sub-atomic scale to generate products with physical and chemical properties that are different to the traditional counterparts [122, 123]. Nanomedicine, on the other hand, is the application of nanotechnology in medicine or health care. Recently, it has become possible to create nanoparticles that can function in an organized manner as biological sensors for the early detection, monitoring and treatment/management of diseases [122]. Nanomedicine has provided novel ways of diagnosis and treatment of dermatological conditions. Although much of the new technologies involving the coupling of nanotechnology with other testing methods are still under investigation, it is a known fact that nanotechnology promises to generate tools and devices that would lead to an increase in the accuracy of detection and analysis of biological samples (e.g. hair). Even though, the use of nanotechnology in trichology is not prevalent, most of the current and non-invasive methods of diagnosis involving the use of nanotechnology may apply to the molecular and structural analysis of hair. Such method that mainly includes imaging and chemical analysis are often referred to as "Nanoimaging" or "Nanoanalytics" [124]. Nanoimaging, which involve the use of nanoparticles to improve the optical and spatial resolution of test materials have proved very useful in diagnosis and testing. For instance, different nanoparticles have been tested in various diagnostic applications due to some advantages such as high sensitivity and specificity, which allow for the use of a little amount biological sample [124–126].

5.2.3. Monitoring trends of/compliance to a nutritional and/or physical activity (PA) protocol

Hair testing can potentially contribute to weight and physical activity monitoring. Exploring this approach would provide a more measurable and objective means to assess an individual's compliance to a reduced calorie, low carb and/or low fat dietary regimens. This is more reliable than what food frequency or PA questionnaires can accurately assess [127]. Post-translationally modified proteins, phase-I/-II enzymes, heat shock proteins, redox status proteins, etc. would be differentially expressed and may better capture dietary/PA compliance than what the classical clinical biochemical parameters (triglycerides, cholesterol, free sugar levels, etc.) would be able to do. These biochemical parameters are rather transitory and are prone to day-to-day confounding factors. For example, specific proteins may serve as excellent surrogate markers of steady-state homeostasis achieved by a more consistent compliance to a dietary or PA regimen.

6. Conclusion

We have discussed various application of non-invasive hair testing in the era of precision medicine and factors that have militated against the routine use of hair testing in diagnostic pathology. Establishing a workflow that can permit the routine use of hair as a testing substrate for disease diagnosis would greatly benefit diagnostic and therapeutic aspects of precision medicine. The use of emerging omics'-based molecular approaches in hair testing would shed more light on hair-based molecular signatures for various physiological and pathological conditions.

Acknowledgements

HAA thanks the South African Medical Research Council (SAMRC) for a mid-career scientist research fellowship. NPK thanks the SAMRC and the National Research Foundation South African Research Chair Initiative.

Conflict of interest

The Authors declare that they have no conflict of financial or non-financial interest.

Author details

Henry Ademola Adeola*, Jennifer Cathryn Van Wyk, Afolake Temitope Arowolo and Nonhlanhla Patience Khumalo

*Address all correspondence to: henry.adeola@uct.ac.za

Hair and Skin Research Laboratory, Division of Dermatology-Department of Medicine, Faculty of Health Sciences, University of Cape Town, South Africa

References

[1] Sachs H. History of hair analysis. Forensic Science International. 1997;**84**(1-3):7-16

[2] Casper J. Practisches Handbuch Der Gerichtlichen Medicin, A. Hirschwald; 1871

[3] Goldblum RW, Goldbaum LR, Piper WN. Barbiturate concentrations in the skin and hair of Guinea pigs. The Journal of Investigative Dermatology. 1954;**22**(2):121-128

[4] Baumgartner AM et al. Radioimmunoassay of hair for determining opiate-abuse histories. Journal of Nuclear Medicine. 1979;**20**(7):748-752

[5] Arnold W. The estimation of medicaments in human hair. Satellite Conference to the 8th Int Conf on Alcohol, Drugs and Traffic Safety, Umeå, Sweden; 1980

[6] Klug E. Zur Morphinbestimmung in kopfhaaren. Zeitschrift für Rechtsmedizin. 1980;**84**: 189-193

[7] Montagna M et al. Simultaneous hair testing for opiates, cocaine, and metabolites by GC-MS: A survey of applicants for driving licenses with a history of drug use. Forensic Science International. 2000;**107**(1-3):157-167

[8] Glaister J. Study of Hairs and Wools Belonging to the Mammalian Group of Animals, Including a Special Study of Human Hair, Considered from the Medico-Legal Aspects; 1931

[9] Hicks J. Microscopy of Hairs: A Practical Guide and Manual. FBI Laboratory: Federal Bureau of Investigation; 1977

[10] Hsu S. FBI Admits Flaws in Hair Analysis over Decades. Post TW, Editor; 2015

[11] Johannessen CM, Clemons PA, Wagner BK. Integrating phenotypic small-molecule profiling and human genetics: The next phase in drug discovery. Trends in Genetics. 2015;**31**(1):16-23

[12] Soderholm S et al. Multi-omics studies towards novel modulators of influenza a virus-host interaction. Viruses. 2016 Sep 29;**8**(10):pii: E269

[13] Hrdy D, Baden HP. Biochemical variation of hair keratins in man nonhuman primates. American Journal of Physical Anthropology. 1973;**39**(1):19-24

[14] Khumalo NP, Gumedze F. African hair length in a school population: A clue to disease pathogenesis? Journal of Cosmetic Dermatology. 2007;**6**(3):144-151

[15] Ebling FJ. The biology of hair. Dermatologic Clinics. 1987;**5**(3):467-481

[16] Shimomura Y, Christiano AM. Biology and genetics of hair. Annual Review of Genomics and Human Genetics. 2010;**11**:109-132

[17] Yang FC, Zhang Y, Rheinstadter MC. The structure of people's hair. PeerJ. 2014;**2**:e619

[18] Robbins CR. Chemical and Physical Behavior of Human Hair. Springer Science & Business Media. Springer-Verlag Berlin Heidelberg; 2012

[19] Orfanos C, Ruska H. Fine structure of human hair. II. Hair-cortex. Archiv fur klinische und experimentelle Dermatologie. 1968;**231**(3):264-278

[20] Wolfram L. The reactivity of human hair. A review. Hair Research. 1981, Springer. pp. 479-500

[21] Cruz CF et al. Keratins and lipids in ethnic hair. International Journal of Cosmetic Science. 2013;**35**(3):244-249

[22] Harkey MR. Anatomy and physiology of hair. Forensic Science International. 1993;**63**(1):9-18

[23] Stenn K. Exogen is an active, separately controlled phase of the hair growth cycle. Journal of the American Academy of Dermatology. 2005;**52**(2):374-375

[24] Van Neste D, Leroy T, Conil S. Exogen hair characterization in human scalp. Skin Research and Technology. 2007;**13**(4):436-443

[25] Henderson G. Mechanisms of drug incorporation into hair. Forensic Science International. 1993;**63**(1):19-29

[26] Namkoong S et al. Reliability on intra-laboratory and inter-laboratory data of hair mineral analysis comparing with blood analysis. Annals of Dermatology. 2013;**25**(1):67-72

[27] Vignali C et al. Hair testing and self-report of cocaine use. Forensic Science International. 2012;**215**(1-3):77-80

[28] Kintz P. Value of the concept of minimal detectable dosage in human hair. Forensic Science International. 2012;**218**(1-3):28-30

[29] Haller DL et al. Hair analysis versus conventional methods of drug testing in substance abusers seeking organ transplantation. American Journal of Transplantation. 2010;**10**(5):1305-1311

[30] Russell E et al. Toward standardization of hair cortisol measurement: Results of the first international interlaboratory round robin. Therapeutic Drug Monitoring. 2015;**37**(1):71-75

[31] Chu TW, Santos L, McElwee KJ. Biology of the hair follicle and mechanisms of nonscarring and scarring alopecia. Seminars in Cutaneous Medicine and Surgery. 2015;**34**(2):50-56

[32] Gandhi M et al. Protease inhibitor levels in hair strongly predict virologic response to treatment. AIDS. 2009;**23**(4):471-478

[33] Hickey. Antiretroviral Concentrations in Small Hair Samples as a Feasible Marker of Adherence in Rural Kenya (vol 66, pg 311, 2014). Jaids-Journal of Acquired Immune Deficiency Syndromes. 2015;**69**(1):E42-E42

[34] Olds PK et al. Assessment of HIV antiretroviral therapy adherence by measuring drug concentrations in hair among children in rural Uganda. Aids Care-Psychological and Socio-Medical Aspects of Aids/Hiv. 2015;**27**(3):327-332

[35] Burhan E et al. Isoniazid, rifampin, and pyrazinamide plasma concentrations in relation to treatment response in Indonesian pulmonary tuberculosis patients. Antimicrobial Agents and Chemotherapy. 2013;**57**(8):3614-3619

[36] Gerona R et al. Quantifying Isoniazid levels in small hair samples: A novel method for assessing adherence during the treatment of latent and active tuberculosis. PLoS One. 2016;**11**(5):e0155887

[37] Mave V et al. Isoniazid hair concentrations in children with tuberculosis: A proof of con cept study. The International Journal of Tuberculosis and Lung Disease. 2016;**20**(6): 844-847

[38] Wright KD, Hickman R, Laudenslager ML. Hair cortisol analysis: A promising biomarker of HPA activation in older adults. Gerontologist. 2015;**55**(Suppl 1):S140-S145

[39] Goldberg SB et al. Hair Cortisol as a biomarker of stress in mindfulness training for smokers. Journal of Alternative and Complementary Medicine. 2014;**20**(8):630-634

[40] Pereg D et al. Hair cortisol and the risk for acute myocardial infarction in adult men. Stress-The International Journal on the Biology of Stress. 2011;**14**(1):73-81

[41] Crunelle CL et al. Hair ethyl glucuronide levels as a marker for alcohol use and abuse: A review of the current state of the art. Drug and Alcohol Dependence. 2014;**134**:1-11

[42] Zimmet P, Alberti KG, Shaw J. Global and societal implications of the diabetes epidemic. Nature. 2001;**414**(6865):782-787

[43] American Diabetes A. Diagnosis and classification of diabetes mellitus. Diabetes Care. 2004;**27**(Suppl 1):S5-S10

[44] Sherif S, Sumpio BE. Economic development and diabetes prevalence in MENA countries: Egypt and Saudi Arabia comparison. World Journal of Diabetes. 2015;**6**(2):304-311

[45] Rosson JW et al. Hair chromium concentrations in adult insulin-treated diabetics. Clinica Chimica Acta. 1979;**93**(3):299-304

[46] Oimomi M et al. Hair protein glycation as a long-term index of blood glucose in diabetics. Diabetes Research and Clinical Practice. 1988;**5**(4):305-308

[47] Kobayashi K, Igimi H. Glycation index of hair for non-invasive estimation of diabetic control. Biological & Pharmaceutical Bulletin. 1996;**19**(4):487-490

[48] Rashaid AH, Harrington Pde B, Jackson GP. Profiling amino acids of Jordanian scalp hair as a tool for diabetes mellitus diagnosis: A pilot study. Analytical Chemistry. 2015;**87**(14):7078-7084

[49] Pichini S et al. Pediatric exposure to drugs of abuse by hair testing: Monitoring 15 years of evolution in Spain. International Journal of Environmental Research and Public Health. 2014;**11**(8):8267-8275

[50] Pumford NR, Halmes NC, Hinson JA. Covalent binding of xenobiotics to specific proteins in the liver. Drug Metabolism Reviews. 1997;**29**(1-2):39-57

[51] Hirakawa T et al. Geranylgeranylacetone induces heat shock proteins in cultured Guinea pig gastric mucosal cells and rat gastric mucosa. Gastroenterology. 1996;**111**(2):345-357

[52] Gelardi A et al. Induction by xenobiotics of phase I and phase II enzyme activities in the human keratinocyte cell line NCTC 2544. Toxicology In Vitro. 2001;**15**(6):701-711

[53] Mnif W et al. Effect of endocrine disruptor pesticides: A review. International Journal of Environmental Research and Public Health. 2011;**8**(6):2265-2303

[54] Blaurock-Busch E, Amin OR, Rabah T. Heavy metals and trace elements in hair and urine of a sample of arab children with autistic spectrum disorder. Maedica (Buchar). 2011;**6**(4):247-257

[55] Smith-Baker C, Saleh MA. Hair as a marker for pesticides exposure. Journal of Environmental Science and Health Part B-Pesticides Food Contaminants and Agricultural Wastes. 2011;**46**(7):648-653

[56] Posecion N Jr et al. Detection of exposure to environmental pesticides during pregnancy by the analysis of maternal hair using GC-MS. Chromatographia. 2006;**64**(11-12):681-687

[57] Toriba A et al. Quantification of polycyclic aromatic hydrocarbons (PAHs) in human hair by HPLC with fluorescence detection: A biological monitoring method to evaluate the exposure to PAHs. Biomedical Chromatography. 2003;**17**(2-3):126-132

[58] Frisch M, Schwartz BS. The pitfalls of hair analysis for toxicants in clinical practice: Three case reports. Environmental Health Perspectives. 2002;**110**(4):433-436

[59] Agrawal M, Biswas A. Molecular diagnostics of neurodegenerative disorders. Frontiers in Molecular Biosciences. 2015;**2**:54

[60] Inamadar AC, Palit A. Neurodegenerative disorders with hair abnormalities: An emergency room consultation for dermatologists. International Journal of Trichology. 2009;**1**(1):30-32

[61] Tamburo E et al. Trace elements in scalp hair samples from patients with relapsing-remitting multiple sclerosis. PLoS One. 2015;**10**(4):e0122142

[62] Chen X et al. Red hair, MC1R variants, and risk for Parkinson's disease–A meta-analysis. Annals of Clinical Translational Neurology. 2017;**4**(3):212-216

[63] Czerny B et al. Screening of trace elements in hair of the female population with different types of cancers in Wielkopolska region of Poland. Scientific World Journal. 2014;**2014**:953181

[64] Blaurock-Busch E et al. Comparing the metal concentration in the hair of cancer patients and healthy people living in the malwa region of Punjab, India. Clinical Medicine Insights: Oncology. 2014;**8**:1-13

[65] Corino GL et al. Characterization of a test for invasive breast cancer using x-ray diffraction of hair-results of a clinical trial. Breast Cancer (Auckl.). 2009;**3**:83-90

[66] Wu X, Tang J, Xie M. Serum and hair zinc levels in breast cancer: A meta-analysis. Scientific Reports. 2015;**5**:12249

[67] Mistry DA, Haklani J, French PW. *Identification of breast cancer-associated lipids in scalp hair*. Breast Cancer (Auckl.). 2012;**6**:113-123

[68] Masukawa Y, Tsujimura H, Narita H. Liquid chromatography-mass spectrometry for comprehensive profiling of ceramide molecules in human hair. Journal of Lipid Research. 2006;**47**(7):1559-1571

[69] Han X, Jiang X. A review of lipidomic technologies applicable to sphingolipidomics and their relevant applications. European Journal of Lipid Science and Technology. 2009;**111**(1):39-52

[70] Yasuda H et al. Metallomics study using hair mineral analysis and multiple logistic regression analysis: Relationship between cancer and minerals. Environmental Health and Preventive Medicine. 2009;**14**(5):261-266

[71] Parker GJ et al. Demonstration of protein-based human identification using the hair shaft proteome. PLoS One. 2016;**11**(9):e0160653

[72] Barthelemy NR et al. Proteomic tools for the investigation of human hair structural proteins and evidence of weakness sites on hair keratin coil segments. Analytical Biochemistry. 2012;**421**(1):43-55

[73] Lee YJ, Rice RH, Lee YM. Proteome analysis of human hair shaft: From protein identification to posttranslational modification. Molecular & Cellular Proteomics. 2006;**5**(5):789-800

[74] Wang L et al. Differential expression of proteins associated with the hair follicle cycle–proteomics and bioinformatics analyses. PLoS One. 2016;**11**(1):e0146791

[75] Laatsch CN et al. Human hair shaft proteomic profiling: Individual differences, site specificity and cuticle analysis. PeerJ. 2014;**2**:e506

[76] Sinclair R et al. The proteomic profile of hair damage. British Journal of Dermatology. 2012;**166**:27-32

[77] Rohner TC, Staab D, Stoeckli M. MALDI mass spectrometric imaging of biological tissue sections. Mechanisms of Ageing and Development. 2005;**126**(1):177-185

[78] Aichler M et al. Spatially resolved quantification of gadolinium(III)-based magnetic resonance agents in tissue by MALDI imaging mass spectrometry after in vivo MRI. Angewandte Chemie (International Ed. in English). 2015;**54**(14):4279-4283

[79] Waki ML et al. Investigation by imaging mass spectrometry of biomarker candidates for aging in the hair cortex. PLoS One. 2011;**6**(10):e26721

[80] Porta T et al. Single hair cocaine consumption monitoring by mass spectrometric imaging. Analytical Chemistry. 2011;**83**(11):4266-4272

[81] Lueking A et al. Profiling of alopecia areata autoantigens based on protein microarray technology. Molecular & Cellular Proteomics. 2005;**4**(9):1382-1390

[82] Liu PR, Regnier FE. Recognizing single amino acid polymorphism in proteins. Analytical Chemistry. 2003;**75**(19):4956-4963

[83] Gomez-Cabrero D et al. Data integration in the era of omics: Current and future challenges. BMC Systems Biology. 2014;**8**(Suppl 2):I1

[84] Uhlen M et al. Transcriptomics resources of human tissues and organs. Molecular Systems Biology. 2016;**12**(4):862

[85] List M et al. KeyPathwayMinerWeb: Online multi-omics network enrichment. Nucleic Acids Research. 2016;**44**(W1):W98-W104

[86] Blumenberg M. SKINOMICS: Transcriptional profiling in dermatology and skin biology. Current Genomics. 2012;**13**(5):363-368

[87] Li S, Ganguli-Indra G, Indra AK. Lipidomic analysis of epidermal lipids: A tool to predict progression of inflammatory skin disease in humans. Expert Review of Proteomics. 2016;**13**(5):451-456

[88] Oliver SG et al. Systematic functional analysis of the yeast genome. Trends in Biotechnology. 1998;**16**(9):373-378

[89] Dettmer K, Hammock BD. Metabolomics–A new exciting field within the "omics" sciences. Environmental Health Perspectives. 2004;**112**(7):A396

[90] Oliver S. Functional genomics: All the King's horses AND all the King's men CAN put humpty together again. Molecular Cell. 2003;**12**(6):1343-1344

[91] Li B et al. Novel applications of metabolomics in personalized medicine: A mini-review. Molecules. 2017;**22**(7):1173

[92] Wikoff WR et al. Pharmacometabolomics reveals racial differences in response to atenolol treatment. PLoS One. 2013;**8**(3):e57639

[93] Cecatti JG et al. Use of metabolomics for the identification and validation of clinical biomarkers for preterm birth: Preterm SAMBA. BMC Pregnancy and Childbirth. 2016;**16**(1):212

[94] Coderch L et al. Lamellar rearrangement of internal lipids from human hair. Chemistry and Physics of Lipids. 2008;**155**(1):1-6

[95] Masukawa Y, Narita H, Imokawa G. Characterization of the lipid composition at the proximal root regions of human hair. International Journal of Cosmetic Science. 2005;**27**(3):191-191

[96] Cornellison C et al. MALDI-MS redox lipidomics applied to human hair: A first look. International Journal of Trichology. 2011;**3**(1):25

[97] Khidhir KG. Human scalp hair follicles express prostaglandin E2 lipid mediator and receptors for Pge2. Science Journal of University of Zakho. 2017;**1**(2):533-542

[98] Lyman DJ, Murray-Wijelath J. Fourier transform infrared attenuated total reflection analysis of human hair: Comparison of hair from breast cancer patients with hair from healthy subjects. Applied Spectroscopy. 2005;**59**(1):26-32

[99] Briki F et al. Breast-cancer diagnosis using hair. Nature. 1999;**400**(6741):226

[100] Weissleder R, Mahmood U. Molecular imaging. Radiology. 2001;**219**(2):316-333

[101] Whitley A, Leroy E, Adar F. FT-IR–Raman Combination: The Perfect Analytical Solution for Vibrational Spectroscopists;2009

[102] Theodorakopoulos N, Chapon V, Coppin F, Floriani M, Vercouter T, Sergeant C, Camilleri V, Berthomieu C, Fevrier L. Use of combined microscopic and spectroscopic techniques to reveal interactions between uranium and Microbacterium sp. A9, a strain isolated from the Chernobyl exclusion zone. Journal of Hazardous Materials. 2015;**285**:285-293

[103] Hendriks R, Lucassen GW. Two-photon fluorescence microscopy of in-vivo human skin. Proceedings of the SPIE. 2000;**4164**:116-121

[104] Konig K. Multiphoton microscopy in life sciences. Journal of Microscopy. 2000;**200**(Pt 2):83-104

[105] Ehlers A et al. Multiphoton fluorescence lifetime imaging of human hair. Microscopy Research and Technique. 2007;**70**(2):154-161

[106] Rompolas P et al. Live imaging of stem cell and progeny behaviour in physiological hair-follicle regeneration. Nature. 2012;**487**(7408):496-499

[107] Shimojo N et al. Mercury dynamics in hair of rats exposed to methylmercury by synchrotron radiation X-ray fluorescence imaging. Life Sciences. 1997;**60**(23):2129-2137

[108] Homma-Takeda S et al. Application of synchrotron radiation X-ray fluorescence imaging combined with histochemical staining to the renal section of mercury-treated rats. Journal of Synchrotron Radiation. 1998;**5**(Pt 1):57-59

[109] Iida A, Noma T. Synchrotron X-ray muprobe and its application to human hair analysis. Nuclear Instruments and Methods in Physics Research Section B: Beam Interactions with Materials and Atoms. 1993;**82**:129-138

[110] Kazarian SG, Chan KL. Applications of ATR-FTIR spectroscopic imaging to biomedical samples. Biochimica et Biophysica Acta. 2006;**1758**(7):858-867

[111] Kazarian SG, Chan KL. ATR-FTIR spectroscopic imaging: Recent advances and applications to biological systems. Analyst. 2013;**138**(7):1940-1951

[112] Chan KL, Kazarian SG. New opportunities in micro- and macro-attenuated total reflection infrared spectroscopic imaging: Spatial resolution and sampling versatility. Applied Spectroscopy. 2003;**57**(4):381-389

[113] Chan KL et al. Fourier transform infrared imaging of human hair with a high spatial resolution without the use of a synchrotron. Applied Spectroscopy. 2005;**59**(2):149-155

[114] Ly E et al. Differential diagnosis of cutaneous carcinomas by infrared spectral micro-imaging combined with pattern recognition. Analyst. 2009;**134**(6):1208-1214

[115] Wang X et al. The comparison of hair from gastric cancer patients and from healthy persons studied by infrared microspectroscopy and imaging using synchrotron radiation. Cancer Epidemiology. 2010;**34**(4):453-456

[116] Zhang GJ, Senak L, Moore DJ. Measuring changes in chemistry, composition, and molecular structure within hair fibers by infrared and Raman spectroscopic imaging. Journal of Biomedical Optics. 2011;**16**(5)

[117] Marcott C et al. Localization of human hair structural lipids using nanoscale infrared spectroscopy and imaging. Applied Spectroscopy. 2014;**68**(5):564-569

[118] Bornschlogl T et al. Keratin network modifications lead to the mechanical stiffening of the hair follicle fiber. Proceedings of the National Academy of Sciences of the United States of America. 2016;**113**(21):5940-5945

[119] Ryals BM, Westbrook EW. Tem analysis of neural terminals on autoradiographically identified regenerated hair-cells. Hearing Research. 1994;**72**(1-2):81-88

[120] Hicks J et al. Uncombable hair (cheveux incoiffables, pili trianguli et canaliculi) syndrome: Brief review and role of scanning electron microscopy in diagnosis. Ultrastructural Pathology. 2001;**25**(2):99-103

[121] Hillmann K, Blume-Peytavi U. Diagnosis of hair disorders. Seminars in Cutaneous Medicine and Surgery. 2009;**28**(1):33-38

[122] Boixeda P et al. Future prospects in dermatologic applications of lasers, nanotechnology, and other new technologies. Actas Dermo-Sifiliográficas. 2015;**106**(3):168-179

[123] Antonio JR et al. Nanotechnology in dermatology. Anais Brasileiros de Dermatologia. 2014;**89**(1):126-136

[124] Iaminskii IV, Gorelkin PV, Dubrovin EV. Nanoanalytics for medicine. Biofizika. 2011;**56**(5):955-960

[125] Papakostas D et al. Nanoparticles in dermatology. Archives of Dermatological Research. 2011;**303**(8):533-550

[126] Kim JK, Nasir A, Nelson KC. Nanotechnology and the Diagnosis of Cutaneous Malignancies. Anonymous Nanotechnology in Dermatology. 2013, Springer: 127-132

[127] Matthews CE et al. Best practices for using physical activity monitors in population-based research. Medicine and Science in Sports and Exercise. 2012;**44**:S68-S76

[128] Yu Y et al. Structure and mechanical behavior of human hair. Materials Science & Engineering. C, Materials for Biological Applications. 2017;**73**:152-163

[129] Everett JR. A New Paradigm for Known Metabolite Identification in Metabonomics/Metabolomics: Metabolite Identification Efficiency. Comput Struct Biotechnol J. 2015;**13**:131-144

Section 4

Industrial Treatment of Keratin

Chapter 8

Keratinaceous Wastes and Their Valorization through Keratinolytic Microorganisms

Debananda Singh Ningthoujam,
Keishing Tamreihao, Saikat Mukherjee,
Rakhi Khunjamayum, Laishram Jaya Devi and
Roshan Singh Asem

Additional information is available at the end of the chapter

http://dx.doi.org/10.5772/intechopen.80051

Abstract

Keratin is a fibrous protein mainly found in higher vertebrates such as mammals, birds, and reptiles. It is also a major constituent of human epithelial tissues. Major keratinaceous wastes include skin, hair, wool, feather, horns, hooves, and nails. Large amounts of such wastes are generated from meat industry, poultry houses, and wool industry etc. Though keratinous wastes contain about 90% protein, keratin is usually recalcitrant to normal proteases. Such wastes have been traditionally digested using physico-chemical methods. But such techniques are energy-intensive and technologically demanding. Also, such approaches lead to degradation of certain amino acids such as lysine. In nature, keratinaceous wastes don't accumulate indicating that keratinolytic microorganisms exist in nature. Keratinase producing strains are distributed among bacteria, fungi, and actinobacteria etc. Hence, potent keratinolytic microbes and their enzymes may be used for valorization of keratinous wastes. Efficient degradation of such wastes may generate value-added products such as feed additives, agricultural biofertilizers, and cosmetics. This chapter will give a comprehensive overview of types of keratinaceous wastes, kinds of keratinolytic microbes and keratinases, and valorization of such wastes using keratinase producing strains and/or keratinases.

Keywords: keratinaceous wastes, keratin, keratinolytic, valorization, keratinase, bacteria, actinobacteria

1. Introduction

Keratin, a versatile bioactive polymer is abundantly distributed in nature. Keratin has turned out to be very attractive for many advanced applications, ranging from agriculture to biomedical engineering. It is the major component of epidermis as found in hair, nail, feather, wool, hooves, scales, and horn. An insoluble macromolecule with long polypeptide chains, keratin is recalcitrant to attack by common proteases like pepsin, papain, and trypsin due to its highly supercoiled and tightly packed molecular structure that is stabilized by inter chain cross linking between parallel strands as well as by hydrogen bonding or hydrophobic interaction [1]. Keratins have been subdivided into α and β [2, 3] based on the secondary structure conformations.

In both α and β configurations, keratin fibrils are twisted parallelly forming micro and macro fibrils conferring stability to the keratin fibers [4, 5]. Keratins can also be subdivided into soft and hard keratins based on the sulfur content.

Hard keratins have high disulfide bond content that makes them tough and inextensible as found in feathers, hair, hooves, and nails. On the contrary, soft keratins are more pliable due to low content of disulfide bonds as found in skin [3, 6]. Properties of any keratinous material such as viscoelasticity and stiffness depend on the degree of hydration of the keratin molecule [7].

Among the keratinous materials, feather is the most predominant one. Millions of tons of feathers are liberated annually from poultry processing farms as waste products and approximately 90% is keratin [8]. Presence of keratin makes chicken feather highly stable environmentally [9]. Feather constitutes 95–98% protein, predominantly β-keratin. The major dominating amino acids in the structure comprise glycine, alanine, serine, cysteine, and valine. The structure has less lysine, methionine, and tryptophan [10].

Apart from feather, keratin is also the major constituent of wool, almost 95% of the dry matter of a wool fiber. The wool fiber is actually a collection of elongated cells that consist of multiple types of keratin proteins. The fiber has three main areas classified into cuticle, cortex, and medulla. The major body of the hair fiber, the cortex, is composed of many spindle shaped cells containing keratin filaments [11].

Apart from feather and wool, mammalian hair also contains keratin [12]. Similarly, hoof horn consists of keratin arranged in both tubular and anti-tubular form. However, in ram horns, fibrous proteins which are alpha keratin in nature and rich in cysteine are found [13].

2. Keratinaceous wastes

Each year millions of tons of keratinous wastes get generated globally especially in wool textile industry and in poultry slaughterhouses [14, 15]. Keratinous wastes, generated mainly in the form of feathers, hairs, horns, hooves, and nails are gradually accumulating in the environment. Enormous amount of urban wastes are accumulating in form of sewage under the bottom sediments of rivers and canals making it difficult to solid waste management and is important to recycle it [9].

Feather is the most abundant keratinous waste material liberated in the modern society. Worldwide annual feather amounts to about 8–9 × 10^5 tons [12]. Chicken meat processing industry contributes maximum to this amount. This industry is growing rapidly as consumption of chicken meat is common to all sections of society, encompassing all customs and religion [16]. According to the USA Foreign Agricultural Service, the total domestic per capita consumption of chickens is 59 kg in the USA; 48.0 kg in Saudi Arabia, 67.1 kg in Hong Kong, 69.7 kg in Israel, and 35.4 kg in Canada [17, 18]. The large consumption of chicken meat generates huge amounts of chicken feathers worldwide.

As per statistics, around 58 × 10^9 chickens are slaughtered for meat in the world every year [18, 19]. The United States of Department of Agriculture figured that 46.6 × 10^9 kg of chicken meat was processed in the USA poultry processing industry in 2014 which generated 40 × 10^9 kg of feathers per annum worldwide [18, 19].

Besides, during poultry processing, many inedible by-products unfit for human consumption are produced. Locations related with animal husbandry and meat products establishments are major sites of environmental pollution and possible transmission of diseases through improper treatments can be possible [20].

Presence of microbial toxins and high quantities of microorganisms, e.g., microbes, infections, parasites and yeasts are common in poultry products [20]. Thus, slaughterhouse products could also be a potential danger to human and expert efforts are needed for management of the recalcitrant keratinous wastes.

3. Keratin digestion

It is estimated that 58 × 10^9 chickens are killed each year which generates enormous keratinous wastes that might create environmental pollution. Poultry processing farms throw around 40 × 10^9 feathers into landfills. Various conventional waste disposal methods such as burial, incineration, and controlled landfilling are practiced. But they have high water and energy demands. Besides, there are also health concerns such as bird flu due to presence of pathogenic microorganisms in dead chicken [21, 22].

The incineration of keratinous wastes release greenhouse gases creating environmental issues. On the other hand, landfilled keratin wastes take a long time to decay and incineration releases greenhouse gases. Also, the costs associated to dispose the feather waste are high as availability of landfill space is reduced. New environmental laws have been developed to enforce generators to deal with environmental wastes in sustainable way. Industries need to recycle, reutilize, minimize, treat, and dispose waste as the last alternative [23]. Therefore, keratin digestion has always been an important issue to maintain a sustained environment.

3.1. Physico-chemical methods

Various physico-chemical methods are being used for decades to digest the keratin thereby mitigating environmental pollution and generating useful resources from keratinous wastes in various aspects.

3.1.1. Hydrothermal method

This process usually employs high steam pressure (10–15 psi) and/or high temperature (80–140°C) in presence of acid or alkali. The process yields water soluble polypeptides, oligopeptides, and even free amino acids. The major drawback of this process is that keratin hydrolysis by hydrothermal method may cause partial or complete destruction of certain amino acids. Besides, it also leads to loss of essential amino acids such as lysine, methionine, and tryptophan. Also, the process leads to formation of non-nutritive amino acids such as lysinoalanine and lanthionine from cystine and lysine respectively [24, 25].

Lysinoalanine is never used at all by animals as a source of lysine [26]. Besides, the hydrolysis of other amino acids is also decreased by excessive steam and heat treatments [24]. In addition, lanthionine content was found to be inversely proportional to the digestible amino acids that suggest presence of lanthionine in feather meal represent an excellent index of over-processing. Another phenomenon which also influences protein quality during this process is the racemization of amino acids. This happens readily after alkaline treatments [27, 28], but to a lesser extent during the heating of proteins [29–31]. Feather meal autoclaved with sodium hydroxide reduced amino acid digestibility values when compared with samples without alkali or enzyme treated.

3.1.2. Acid and alkaline treatment

The pros and cons of acid and alkaline hydrolysis of keratinous wastes have been described by Asquith [32]. The most satisfactory method to convert keratin quantitatively into their individual amino acids involves acid hydrolysis [33, 34]. Martin and Synge examined partial acid hydrolysis of wool and gelatin at 37°C in an excess of 10 N HCI for several days and were succeeded in release of one third of amino acids [35, 36]. Asquith was able to semi-quantitatively determine some peptide sequences in wool keratin by controlled hydrolysis of keratose fractions [37]. Similarly, partial hydrolysis was used to determine the amino acid sequence of wool proteins [38, 39].

Alkali also hydrolyzes keratin fibers but less selectively than to acids. 0.1 N NaOH rapidly dissolves wool while boiling. However, the process results in destruction of arginine, serine, threonine, cystine, and cysteine. But tryptophan is not destroyed in alkali, and the analysis of alkaline hydrolysates is done by quantitative determination of tryptophan [40]. Wool, if treated with alkali, three new amino acids- Ianthionine, lysinoalanine, and 8-aminoalanine could be recovered [40].

3.1.3. Steam explosion

Steam explosion is a subtype of hydrothermal method of keratin hydrolysis and have been discussed by various authors [41]. Steam explosion (SE), a hydroelectric pre-treatment of biomass releases the constitutive components. The process involves short exposure and then rapid release of the pressure in an explosive decompression event [42]. Originally developed by Mason, the process has been explored extensively in biomass conversion [43, 44].

Using this methodology, Tonin et al. [45] generated hydrolysate from wool waste. But the yield was 18.66%.

The hydrolysis of wool wastes by steam explosion was also studied by Xu et al., 2006 [46]. A yield of 62.5% hydrolysates was achieved by passing steam at 600°C and 0.8 MPa. Scanning Electron Microscopy results detected that during explosion, some scales on the fiber surface were cleaved and tiny grooves were formed. Differential Scanning Electron Microscopy indicates reduction in thermal decomposition energy of the treated fiber. This reflects destruction of crystals and crosslinks of macromolecular chains in the fiber due to steam explosion. Besides, steam explosion also did some alterations in the fiber properties such as reduction in strength, solubility in caustic solution, and moisture regain.

3.1.4. Ionic liquids

Use of ionic liquids such as 1-butyl-3-methylimidazolium chloride ([BMIM]Cl) or a hydrophobic ionic liquid (IL), 1-hydroxyethyl-3-methylimidazolium bis(trifluoromethanesulfonyl) amide ([HOEMIm][NTf2]) have been attempted to generate keratin hydrolysates [47–49]. The extracted keratin could also be easily separated from the reaction system. Another ionic liquid, NaHSO3, was also successfully investigated. The results indicated [BMIM] Cl/[HOEMIm][NTf2] is an efficient catalyst and solvent for dissolving feathers and could be easily recovered due to its hydrophobicity [47]. The dissolution and regeneration of the waste chicken feathers in an ionic liquid of [BMIM]Cl showed an excellent efficiency (63.5–87.7%).

3.1.5. Reduction and oxidation

Reduction and oxidation have also been used to convert keratinous wastes esp. hair wastes to hydrolysates. Shindai method is a special type of treatment employed for preparing hydrolysates from hair wastes [50, 51]. It is a rapid and convenient procedure to extract human hair proteins for examining the biochemical properties in detail.

The procedure is based on the principle that in the presence of a reductant, a combination of thiourea and urea can effectively remove proteins from the cortex part of human hair. Using this procedure, the extracted fraction from human hair mainly consisted of hard α-keratins with molecular masses of 40–60 kDa, and keratin-associated proteins (KAPs) with a molecular mass of 6–30 kDa. Hair samples when incubated in the Shindai solution containing alcohols such as methanol, ethanol, 1-propanol, 2-propanol, 1-butanol, and 2-methyl-1-propanol, the extraction of KAPs was enhanced, but extraction of keratin was suppressed. Thus by using ethanol, selective purification of KAPs and keratin was achieved.

3.2. Microbial keratin degradation

Microbial degradation of keratin is reported in some bacteria, actinomycetes, keratinophilic fungi, and larvae of the common clothes moth *Tineola bisselliella* [52, 53]. They use keratin as the sole source of C, N, S, and energy. Keratinolytic bacteria were most isolated from bird feathers and the plumage [54–56], composting [57], or feather waste processed by fermentation. The bacteria often belong to the genus Bacillus and order actinomycetes. Feather degrading abilities were mostly found in *Bacillus licheniformis* [54] as well as in *Bacillus pumilis, Bacillus subtilis,* and *Bacillus cereus* [57] and in some non-spore forming bacteria *Stenotrophomonas sp, Fervidobacterium pannavorans* [58], and *F. islandicum* [59]. Some species of actinomycetes intensively degrade keratins such as *Streptomyces* which includes *S. fradiae* [60], *S. pactum* [61] *S. thermoviolaceus* [62], or other actinomycetes such as *Thermoactinomyces* [63].

Keratin degrading activity is also observed in nutrient-specialized keratinophilic fungi, which uses keratin as nutrient [61]. These types of fungi besides colonizing bird plumage and mammal hair, also colonizes natural habitats where keratin material is available such as places inhabited by birds, humans, and mammals [64, 65].

Fungi exhibiting high keratinolytic activities include the following genera: *Aspergillus, Chrysosporium, Alternaria, Trichuris, Monodictys, Myrothecium, Paecilomyces, Stachybotrys, Urocladium, Scopulariopsis, Curvularia, Cladosporium, Fusarium, Geomyces, Gleomastis, Penicillium* and *Doratomyces* [66–68].

Based on the keratinolytic efficiencies, microbes can be divided in two types: true keratinolytic microbes solubilizing hard keratin structures or potentially keratinolytic microbes which have strong proteolytic effects thereby solubilizing non-keratin proteins which are associated with hard keratins in keratin rich material like feathers, nail etc. They can also degrade soft keratins as found in callus [64].

3.2.1. Microbial keratinase

Keratinolytic microbes possess proteolytic enzymes which are able to degrade the extensive disulfide crosslinking of keratin polypeptides and solubilize the keratin. These keratinolytic proteases are known as keratinases (EC 3.4.21/24/99.11). Keratinases have unique capacity to act on compact substrates such as keratinous wastes compared to traditional proteases [69].

Keratin wastes can be efficiently degraded by bacteria, actinomycetes, and fungi due to presence of the keratinases [70].

The keratinases convert feather keratins into fertilizers and useful feedstuffs [69, 71]. Keratinases are primarily extracellular and they show activity on keratinous substances [72]. However, they are also secreted constitutionally in absence of keratin [73].

Some keratinolytic enzymes are also intracellular [71]. Microbial keratinolytic proteases are mostly serine proteinases and sometimes metalloproteinases, being inhibited by phenyl methane sulfonyl fluoride (PMSF) and other inhibitors of serine proteases [74, 75]. The optimum activity of keratinases range between pH 6.0 and 9.0. The enzymes are either neutral or alkaline proteases requiring Ca^{++}, as the activity is inhibited in presence of Ca^{++} chelating EDTA or EGTA [76]. Keratinases are able to hydrolyze both soluble proteins as well as insoluble, fibrous proteins [77].

Although microorganisms capable of degrading keratinous substrates are generally isolated from soil and poultry wastes, these microorganisms are almost ubiquitous in nature, thriving under diverse ecological and environmental conditions [78].

No keratinases can completely solubilize native keratin. However, keratinase contribute to the valorization of the enormous keratin containing wastes in the form of hair, feathers, dead birds, and animals [70]. These enzymes have gained increasing attention due to their biotechnological applications on valorization of keratinous wastes especially byproducts of agro-industrial processes [79, 80]. These enzymes can also be employed for degradation of prions [81, 82] and β-amyloid fibers [83].

3.3. Alkaline-enzymatic methods

Mokrejs et al., 2010 proposed a two-stage alkaline & enzymatic method to hydrolyse keratinous wastes [84]. It involves initial treatment of wastes with 0.1 or 0.3% KOH at 70°C (1:50 ratio) for 24 h. After adjusting the pH to 9.0, enzymatic hydrolysis is carried out by a keratinase (1–5%), which hydrolyzes the wastes further at 50–70°C for 4–8 h.

This two-stage hydrolysis was applied for processing waste chicken feathers. The amount of hydrolyzed feathers after second stage hydrolysis increased to 90.8% achieving high efficiency under mild reaction conditions leading to the generation of keratin hydrolysate.

4. Valorization of keratinous wastes

Microbial keratinases are gradually getting importance biotechnologically for valorization of keratinous wastes. Microbial keratinases are being used successfully in degrading keratin into economically important keratin protein hydrolysates which can find potential applications as animal feed supplements, bio- fertilizers, biodegradable glues, films, and foils [70, 85]. Also, valorization of keratinous waste by keratinolysis finds useful applications in various industries such as elimination of horny epithelial cells that adheres to textile fibers (Textile Industry), clearing obstructions in sewage systems (Waste Water Management Industry), conversion of poultry or agro-industrial wastes into valuable protein products such as amino acids for livestock feed, pharmaceutical, and cosmetic industries. The details of valorization of keratinous wastes are described below.

4.1. Agricultural products

Organic fertilizers have been prepared from sulfur containing amino acids which are richly present in keratin hydrolysates. Chicken feather composts have been reported to act as biofertilizers [55]. Feathers have a high content of nitrogen (13%) and can serve as an excellent compost material or biofertilizers; however, degradation of feather is difficult due to presence of disulfide bonds [86].

Ichida et al., 2001 used compatible strains of *Streptomyces* and *B. licheniformis* as mixed starter culture for inoculating waste feathers in bioreaction vessels. The trials demonstrated bioreactors of compost materials consisting of chicken feathers, straw, and poultry litter while inoculated with feather-degrading bacteria, increased the rate of keratin utilization. Composting is successful if the mixture of organic materials consists of 20 to 40 parts of carbon to 1 part of nitrogen as required by the composting bacteria, actinomycetes and fungi [87].

Besides, feather hydrolysates are promising slow release nitrogen (N) fertilizers [88, 89]. Release of N from feathers has been enhanced by treatment with several keratinolytic bacteria, actinobacteria, and fungi. *Chryseobacterium spp.* is successfully used for preparing slow release N fertilizers from feathers. Such feather-based biofertilizers were found to be effective fertilizers for banana and other crops [63]. Also, feather hydrolysates prepared using thermophilic actinomycetes was successfully applied as fertilizers for ryegrass cultivation. Similarly, Choi & Nelson reported the preparation of slow release nitrogen fertilizer from poultry feather by biodegradation [90].

4.2. Pharmaceutical applications

Chicken feather cholesterol may serve as precursor for bile salt synthesis. Bile salts are used as bio-emulsifiers and biosurfactants in cosmetic industry. Besides, cholesterol is required for synthesis of other pharmaceuticals such as vitamin D3 and steroids [91]. Vitamin D3 is necessary for bone and teeth formation [92].

Keratin has properties suitable for use in biomedical applications like biodegradability, biocompatibility, sterilizability, bioresorbability, functionality, self-assembly, and manufacturability including mechanical and thermal properties [93]. Wool and human hair keratin have been used to prepare protein films, fibers, and scaffolds for tissue engineering. The propensity of extracted keratin to self-aggregate and form 3D structures has enabled it to be used as scaffolds for tissue engineering [94]. Chicken feather keratin can be fabricated into keratin films and the construct can be used in controlled drug delivery [95, 96].

Sun et al., 2009 reported development of keratin microparticles by treating keratin with ionic liquid [48]. Nanoparticles prepared from feather keratin exhibited good biocompatibility and stability thereby opening up the possibility of controlled drug delivery [97]. Such keratin based film generation for controlled drug delivery was also reported [63]. Besides, keratin nanofibers developed by electrospinning is being applied in tissue engineering and regenerative medicine [98, 99].

In addition, keratin biomaterials from chicken feathers are found to be capable of supporting cellular attachment as they possess cell binding motifs, such as glutamic acid-aspartic acid-serine and leucine-aspartic acid-valine binding residues [18].

Recently, human hair keratin has been developed into biomaterials for use in tissue engineering [100, 101].

4.3. Bioactive peptides

Keratinous wastes can also be hydrolyzed to produce bioactive peptides with potential health benefits. Fakhfakh et al., 2012 reported production of keratin hydrolysates containing bioactive peptides with antioxidant activity [102]. Similarly, other authors have studied angiotensin-converting enzyme (ACE)-inhibitory and dipetidyl peptidase-IV (DPP-IV)-inhibitory activities of keratin hydrolysates [103].

These activities find therapeutic applications in medical conditions such as high blood pressure and inflammation.

4.4. Industry

4.4.1. *Livestock industry*

Chicken feather hydrolysates are rich in amino acids and peptides which are of similar composition with that of soybean and cotton seed extracts. Hence, such feather hydrolysates have found promising applications as animal feed additive [104]. Enrichment with lysine may enhance the nutritive value of such feed additives [105].

Horn meal prepared from raw horns and hooves have also been proposed as animal feed additive. Brandelli et al., 2015 have also proposed keratin hydrolysates as promising animal feed [106].

4.4.2. *Cosmetic industry*

Keratin hydrolysates may find tremendous applications in the cosmetic industries. Keratin based cosmetics have been reported as therapeutic agents for skin and human hair. Keratin has been also used as components of blends [107].

Keratin has been reported as components of cosmetic blends along with other natural polymers such as collagen, chitosan, and silk fibroin etc. Keratin or keratin hydrolysates help retaining moisture in the skin by interactions of stratum corneum and hair cuticle with the cosmetics. Also, keratin protects the cortex of the human cell from daily injuries of heat or chemicals. Besides, hydrolyzed keratin is used as a cosmetic ingredient. Topical application of hydrolyzed keratin enhances skin hydration and flexibility [108].

Because of moisture retaining properties, keratin is an important component of shampoos and conditioners, hair loss concealing products, and other hair beautification accessories [109]. The protein hydrolysates provide advantages to the hair by strengthening hair fibers and decreasing fiber breakages. Many plant and animal hydrolysates such as wheat protein, wool, nails, and horns keratin [108, 109] have been used in cosmetic industry as components of hair shading splashes and toners to enhance uniform color retention of hair. Besides, protein hydrolysates act as restorers in hare care processing industry [110].

4.4.3. *Automobile industry*

Automobile industry is planning to use feathers to produce composite materials that can be used in dashboards, seats, car parts, cushioning, and interior lines [18]. Currently petroleum based raw materials are used in automobile and aeroplane parts. However, material science industries prefer low cost, lightweight, and environmentally sustainable materials. Keratins being low priced and light weight as well as possessing enormous strength due to presence of high cysteine content is a choice by engineers of automobile industry.

4.4.4. *Leather and textile industry*

Keratinous waste materials have a promising potential in textile industry. Scientists are investigating the potential of chicken feathers for replacing natural fibers and man-made fibers thereby saving trees in textile processing. Chicken feathers have toughness, flexibility, high surface area, fine diameter, and durability rendering them valuable for replacing natural and synthetic fibers, and wood pulp. However, pre-treatment of feather is important before usage into textile products. The barbs must be stripped off as the barb material of feather shows similar property of a textile fiber [18, 111]. Since feathers contain high nitrogen content, they can be used as flame retardants. Guan and Chen, 2006 reported preparation of hydrolyzed feathers as flame retardant finish [112]. Cotton fibers while treated with this feather-made flame retardant, acquired property of flame retardance.

Chicken feather can also be a good source as textile binding and textile sizing agent in textile printing as keratin has film forming and binding ability. So far, starch and starch derivatives and polyvinyl alcohols were being used as a sizing agent to coat a protective layer on the surface of yarns to improve weaving [113, 114]. Sizing provides tensile strength and abrasion resistance of yarns. But usage of starch is associated with socio-economic problems and polyvinyl alcohol has high cost and poor biodegradability. Therefore, chicken feather could be a promising substitute for starch and polyvinyl alcohol [114].

Similarly, keratin hydrolysates are used in leather industry for usage in filling and retaining. Many processes in leather tanning can cause serious health hazards such as skin and respiratory ailments, including cancer. Recently, Wool has developed biocomposites based on the techniques by aerospace engineers to convert scraped chicken feathers into synthetic leather [18, 115].

4.4.5. Construction industry

Nowadays, composite materials from thermoplastic and natural fibers are being investigated in the construction industry. The natural fibers are mostly cellulosic and provide enormous strength. Usage of natural fibers reduce consumption of synthetic polymers thereby consumption of petroleum products is reduced. However, cellulosic natural fibers are not compatible with the hydrophobic polymer material and chicken feather can be a useful alternative. After separation of feather in various parts such as long fibers, short fibers, and powdered rachis, feathers are easy to melt to be used as reinforced matrix material [18, 114]. Feathers may be valorized to thermoplastic films for use as packaging materials and in other applications. Grafting of feather keratin with acrylic monomers have been shown to improve the thermoplastic properties of feather based materials [116].

Such composites are suggested for usage in thermal and sound insulation as well as ceiling applications. Chicken feather composite boards are being hypothesized to be the substitute for wood and plastic in the construction industry [18].

4.4.6. Environmental remediation

Although being an environmental pollutant, keratinous material, after valorization is commercially used as agents for environmental remediation. Keratinous material has been suggested to be used as environment-friendly electrode materials [117]. Keratin extracted from feather may be diluted, followed by lyophilization, to generate sponges. Such sponges can be useful agents for clean-up of oil spills. Keratin fibers have promise to be developed into innovative green materials due to their biodegradability, biocompatibility, natural abundance, and mechanical durability [118, 119]. Feather keratins are highly hygroscopic and possess excellent absorption ability. Therefore feathers can be used as efficient purification agents for waste water by removal of heavy metals such as Cu, Se, Zn, and other toxic compounds [120, 121].

4.4.7. Energy sector

Keratinous wastes have been used to generate bio-hydrogen [122]. Initially keratinous wastes are converted into a fermentation product rich in amino acids and peptides. Next, minerals are added to the product to serve as substitute for bacto-peptone. The enriched fermentation product is then subjected to further anaerobic fermentation by a thermophilic archaea *Thermococcus sp.* to generate bio-hydrogen.

Also, chicken feathers contain good amount of fat. The fats after extracting from feather meal by solvent extraction can subsequently be transesterified to biodiesel using catalysts. It is estimated that hundreds of millions of liters of biodiesel can be generated from chicken feather waste globally. This would reduce the petroleum dependency as well as cut the carbon emissions [18].

5. Conclusions and future prospects

Published literature shows keratin is a raw material for production of a diversity of value added products. Huge amounts of waste keratinous biomass generated by food, wool, and livestock industries may be utilized gainfully as feedstocks/raw materials for production of keratin and keratin hydrolysates at industrial scales. Valorization of keratinous wastes will not only generate many commercial products but will also ameliorate the environmental pollution from such wastes, and also help boost pharmaceutical, food, and cosmetic industries.

Keratinolytic enzymes will be better alternatives for digestion of keratinous wastes as compared to physico-chemical methods. This dictates search for more efficient keratinolytic bacteria and their enzymes.

Use of keratinous wastes for production of biofertilizers is one of the most recent applications of such waste biomass. Such biofertilizers will be better alternatives to fossil fuel based synthetic fertilizers and will help mitigate climate change.

There is urgent need for developing better keratinolytic strains, characterizing better keratinases, improve methods for generating keratin hydrolysates, and finding other novel applications of keratinaceous wastes.

Research in this area has a promising future.

Acknowledgements

The authors are grateful to Department of Biotechnology, Govt. of India for sponsoring the host laboratory under the project "Establishment of State Biotech Hubs" by order no: *BT/04/NE/2009* and upgrading the project to *Advanced Level State Biotech Hub* (AdL-SBTHub), Manipur University, India.

Conflicts of interests

The authors declare no competing interests.

Author details

Debananda Singh Ningthoujam*, Keishing Tamreihao, Saikat Mukherjee, Rakhi Khunjamayum, Laishram Jaya Devi and Roshan Singh Asem

*Address all correspondence to: debananda.ningthoujam@gmail.com

Advanced State Level Biotech Hub (AdL SBTHub), Department of Biochemistry, Manipur University, Canchipur, Manipur, India

References

[1] Katarzyna C, Helena G, Izabela M, Henryk G. Valorization of keratinous materials a review. Waste and Biomass Valorization. 2011;**2**:317-321

[2] Akhtar W, Edwards HGM. Fourier-transform Raman spectroscopy of mammalian and avian keratotic biopolymers. Spectrochimica Acta. 1997;**53**:81-90

[3] Voet D, Voet JG. Biochemistry. New York: Wiley; 1995. pp. 154-156

[4] Kreplak L, Doucet J, Dumas P, Briki F. New aspects of the α-helix to β- sheet transition in stretched hard α-keratin fibers. Biophysical Journal. 2004;**87**:640-647

[5] Zerdani I, Faid M, Malki A. Feather wastes digestion by new isolated strains *Bacillus sp.* in Morocco. African Journal of Biotechnology. 2004;**3**:67-70

[6] Schrooyen PMM, Dijkstra PJ, Oberthur RC, Bantjes A, Feijen J. Partially carboxymethylated feather keratins. Journal of Agricultural and Food Chemistry. 2001;**49**:221-230S

[7] Collins SN, Cope BC, Hopegood L, Latham RJ, Linford RG, Reilly JD. Stiffness as a function of moisture content in natural materials: Characterisation of hoof horn samples. Journal of Materials Science. 1998;**33**:5185-5191

[8] Santos RMDB, Firmino AAP, Sa CM, Felix CR. Keratinolytic activity of *Aspergillus fumigatus* Fersenius. Current Microbiology. 1996;**33**:364-370

[9] Itisha S, Kushwaha RKS. Keratinases and microbial degradation of keratin. Advances in Applied Science Research. 2015;**6**(2):74-82

[10] Wojciech Ł, Barbara Ż, Dorota C, Anna P, Michał P, Anna K, Wiesław K. New keratinolytic bacteria in valorization of chicken feather waste. AMB Express. 2018;**8**:9

[11] Ingrid-Iona B. Sustainable alternatives for wool valorization. Annals of the University of Oradea: Fascicle of Textiles, Leatherwork. 2014;**XV**(2):27-32

[12] Grazziotin A, Pimentel FA, de Jong EV, Brandelli A. Nutritional improvement of feather protein by treatment with microbial keratinase. Animal Feed Science and Technology. 2006;**126**:135-144

[13] McLean CM, Koller CE, Rodger JC, MacFarlance GR. Mammalian hair as an accumulative bioindicator of metal bioavailability in Australian terrestrial environments. Science of the Total Environment. 2009;**407**:3588-3596

[14] Moreira FG, de Souza CGM, Costa MAF, Reis S, Peralta RM. Degradation of keratinous materials by the plant pathogenic fungus *Myrothecium verrucaria*. Mycopathologia. 2007;**163**:153-160

[15] Aluigi A, Zoccola M, Vineis C, Tonin C, Ferrero F, Canetti M. Study on the structure and properties of wool keratin regenerated from formic acid. International Journal of Biological Macromolecules. 2007;**41**:266-273

[16] Rahayu S, Bata M. Quality of chicken feather processed in different conditions. Animal Production. 2015;**16**(3):170-175

[17] Evans T. Global poultry trends 2014: Slowing of Asian population growth to impact chicken consumption. The Poultry Site. 2014

[18] Tesfaye T, Sithole B, Ramjugernath D. Valorisation of chicken feathers: A review on recycling and recovery route-current status and future prospects. Clean Technologies and Environmental Policy. 2017;**19**(10):2363-2368

[19] Farm Welfare Compendium. Statistics: Broiler Chicken. Compassion in World Farming; FAO stat. 2013. pp. 1-11

[20] Franke-Whittle IH, Insam H. Treatment alternatives of slaughterhouse wastes, and their effect on the inactivation of different pathogens: A review. Critical Reviews in Microbiology. 2013;**39**(2):139-151

[21] Edwards D, Daniel T. Environmental impacts of on-farm poultry waste disposal—A review. Bioresource Technology. 1992;**41**(1):9-33

[22] Urlings H, Van LJ, Bijker P. Slaughter by-products: Problems, preliminary research and possible solutions. Vet Q. 1992;**14**(1):34-38

[23] Molapo NA. Waste handling practices in the South African high throughput poultry abattoirs [Dissertation]. Free State: Central University of Technology; 2009. pp. 21-120

[24] Papadopoulos MC. Effect of processing on high-protein feedstuffs: A review. Biological Wastes. 1989;**29**:123-138

[25] Latshaw JD, Musharaf N, Retrum R. Processing of feather meal to maximize its nutritional value for poultry. Animal Feed Science and Technology. 1994;**47**:179-188

[26] Robbins KR, Baker DH, Finley JW. Studies on the utilization of lysinoalanine and lanthionine. The Journal of Nutrition. 1980;**110**(5):907-915

[27] Provansal MMP, Cug JLA, Cheftel JC. Chemical and nutritional modifications of sunflower proteins due to alkaline processing. Formation of amino acid cross-links and isomerization of lysine residues. Journal of Agricultural and Food Chemistry. 1975;**23**:938-943

[28] Masters PM, Friedman M. Amino acid racemization in alkali-treated food proteins-chemistry, toxicology, and nutritional consequences. In chemical deterioration of proteins. ACS Symposium Series. 1980;**123**:165-194

[29] Sternberg M, Kim CY, Schwende FJ. Lysinoalanine: Presence in foods and food ingredients. Science. 1975;**190**:992-994

[30] Liardon R, Hurrell RF. Amino acid racemization in heated and alkali treated proteins. Journal of Agricultural and Food Chemistry. 1983;**31**:432-437

[31] Schwass DE, Finley JW. Heat and alkaline change to proteins: Racemization and lysinoalanine formation. Journal of Agricultural and Food Chemistry. 1984;**32**:1377-1382

[32] Asquith RS. Chemistry of Natural Protein Fiber. New York: Plenum; 1977. pp. 1-417

[33] Hill RL. Hydrolysis of proteins. Advances in Protein Chemistry. 1965;**20**:37

[34] Light A. Partial acid hydrolysis. Methods in Enzymology. 1967;**11**:417-420

[35] Martin RB, Synge RLM. A new form of chromatogram employing two liquid phases. The Biochemical Journal. 1941;**35**:1358

[36] Asquith RS, Parkinson DC. Proceedings of the Society for Analytical Chemistry. 1968;**5**:206

[37] Asquith RS, Shaw T. Makromolekulare Chemie. 1968;**115**:198

[38] Haylett T, Swart LS. Studies on the high-sulfur proteins of reduced merino wool. Textile Research Journal. 1969;**39**:917

[39] Corfield MC, Fletcher JC. Amino acid sequences of peptides from a chymotryptic digest of a urea soluble protein fraction (U.S.3) from oxidized wool. Biochemistry. 1969;**115**:323

[40] Corfield MC, Robson A. The amino acid composition of wool. The Biochemical Journal. 1955;**59**:62

[41] Yu Z, Bailiang Z, Yu F, Xu G, Andong S. A real explosion: The requirement of steam explosion pretreatment. Bioresource Technology. 2012;**121**:335-341

[42] Glasser WG, Wright RS. Steam-assisted biomass fractionation. II. Fractionation behavior of various biomass resources. Biomass and Bioenergy. 1998;**14**:219-235

[43] Lipinsky ES. Chemicals from biomass: Petrochemical substitution options. Science. 1981;**212**:1465-1471

[44] Bungay HR. Biomass refining. Science. 1982;**218**:643-646

[45] Tonin C, Zoccola M, Aluigi A, Varesano A, Montarsolo A, Vineis C, Zimbardi F. Study on the conversion of wool keratin by steam explosion. Biomacromolecules. 2006;**7**:3499-3504

[46] Xu W, Ke G, Wu J, Wang X. Modification of wool fiber using explosion. European Polymer Journal. 2006;**42**:2168-2173

[47] Wang Y-X, Cao X-J. Extracting keratin from chicken feathers by using hydrophobic ionic liquid. Process Biochemistry. 2012;**47**:896-899

[48] Sun P, Liu Z-T, Liu Z-W. Particles from bird feather: A novel application of an ionic liquid and waste resource. Journal of Hazardous Materials. 2009;**170**:786-790

[49] Ji Y, Chen J, Lv J, Li Z, Xing L, Ding S. Extraction of keratin with ionic liquids from poultry feather. Separation and Purification Technology. 2014;**132**:577-583

[50] Nakamura A, Arimoto M, Takeuchi K, Fujii T. A rapid extraction procedure of human hair proteins and identification of phosphorylated species. Biological & Pharmaceutical Bulletin. 2002;**25**:569-572

[51] Fujii T, Takayama S, Ito Y. A novel purification procedure for keratin-associated proteins and keratin from human hair. International Journal of Biological Macromolecules. 2013;**13**:92-106

[52] Calin M, Aruxandei DC, Alexandresca E, Raut I, Doni MB, Arsune ML, Oncae F, Jeca L, Lazar V. Degradation of keratinous substrates by keratinolytic fungi. Electronic Journal of Biotechnology. 2017;**28**:101-111

[53] Christeller JT, Markwick NP, Burgess EPJ. Midgut proteinase activity of three keratinolytic larvae *Hoffmannophila pseudospreletta*, *Tineola bisselliella*, and *Anthrenocerus australis*, and the effect of proteinase inhibitors on proteolysis. Archives of Insect Biochemistry and Physiology. 1994;**25**(2):159-173

[54] Burtt EH Jr, Ichida JM. Occurrence of feather-degrading bacilli in the plumage of birds. The Auk. 1999;**116**:364-372

[55] Ichida JM, Krizova L, LeFevre CA, Keener HM, Elwell DL, Burtt EH Jr. Bacterial inoculum enhances keratin degradation and biofilms formation in poultry compost. Journal of Microbiological Methods. 2001;**47**:199-208

[56] Lateef A, Oloke J, Kana EBG, Bello BY. Keratinolytic activities of a new feather degrading isolate of *Bacillus cereus* LAU08 isolated from Nigerian soil. International Biodeterioration & Biodegradation. 2010;**64**(2):162-165

[57] Kim JM, Lim WJ, Suh HJ. Feather-degrading *Bacillus* species from poultry waste. Process Biochemistry. 2001;**37**:287-291

[58] Friedrich AB, Antranikian G. Keratin degradation by *Fervidobacterium pennavorans*, novel thermophilic anaerobic species of the order Thermotogales. Applied and Environmental Microbiology. 1996;**62**:2875-2882

[59] Nam G-W, Letourneau D-W, Lin HS, Lee NJ, Kim BC, Choe EA, Hwang JK, Suhartono MT, Pyun YR. Native-feather degradation by *Fervidobacterium islandicum* AW-1, a newly isolated keratinase-producing thermophilic anaerobe. Archives of Microbiology. 2002;**178**:538-547

[60] Noval JJ, Nickerson WJ. Decomposition of native keratin by *Streptomyces fradiae*. Journal of Bacteriology. 1959;**77**:251-225

[61] Bockle B, Galunsky B, Muller R. Characterization of keratinolytic serine protease from *Streptomyces pactum* DSM 40530. Applied and Environmental Microbiology. 1995;**61**: 3705-3710

[62] Chitte RR, Nalawade VK, Dey S. Keratinolytic activity from the broth of a feather-degrading thermophilic *Streptomyces thermoviolaceus* strain SD8. Letters in Applied Microbiology. 1999;**28**:131-136

[63] Gousterova A, Nustorova M, Paskaleva D, Naydenov M, Neshev G, Vasileva-Tonkova E. Assessment of feather hydrolysate from thermophilic actinomycetes for soil amendment and biological control application. International Journal of Environmental Research. 2011;**5**:1065-1070

[64] Korniłłowicz-Kowalska T. Studies on the decomposition of keratin wastes by saprophytic microfungi. I. Criteria for evaluating keratinolytic activity. Acta Mycologica. 1997;**32**:51-79

[65] Pugh GJF. Cellulolytic and keratinophilic fungi recorded on birds. Sabouraudia. 1965;**4**:85-91

[66] Farag AM, Hassan MA. Purification, characterization, immobilization of a keratinase from *Aspergillus oryzae*. Enzyme and Microbial Technology. 2004;**34**:85-93

[67] Gradisar H, Kern S, Friedrich J. Keratinase of *Doratomyces* Microsporus. Applied Microbiology and Biotechnology. 2000;**53**:196-200

[68] Marcondes NR, Taira CL, Vandresan DC, Svidzinski TIE, Kadowaki MK, Peralta RM. New feather–degrading filamentous fungi. Microbial Ecology. 2008;**56**:13-17

[69] Brandelli A. Bacterial keratinases: Useful enzymes for bioprocessing agroindustrial wastes and beyond. Food and Bioprocess Technology. 2008;**1**:105-116

[70] Onifade AA, Al-Sane NA, Al-Musallam AA, Al-Zarban S. Potentials for biotechnological applications of keratin-degrading microorganisms and their enzymes for nutritional improvement of feathers and other keratins as livestock feed resources. Bioresource Technology. 1998;**66**:1-11

[71] Gupta R, Ramnani P. Microbial keratinases and their prospective applications: An overview. Applied Microbiology and Biotechnology. 2006;**70**:21-33

[72] Asahi MR, Lindquist K, Fukuyama G, Apocarda W, Epstein WL, McKerrow JH. Purification and characterization of major extracellular proteinases from *Trichophyton rubrum*. The Biochemical Journal. 1985;**232**:139-144

[73] Sanyal AK, Das SK, Banerjee AB. Purification and partial characterization of an exocellular proteinase from *Trichophyton rubrum*. Sabouraudia. 1985;**23**:165-178

[74] Jaouadi NZ, Rekik H, Badis A, Trabelsi S, Belhoul M, Yahiaou AB, et al. Biochemical and molecular characterization of a serine Keratinase from *Brevibacillus brevis* US575 with promising keratin-biodegradation and Hide-dehairing activities. PLoS One. 2013;**8**(10): e76722

[75] Asker MMS, Mahmoud MG, Shebyi KE, Aziz MSA, Mohamed S, Aziz A. Purification and characterization of two thermostable protease fractions from *Bacillus megaterium*. Journal, Genetic Engineering & Biotechnology. 2013;**11**(2):103-109

[76] Bach E, Sant V, Daniel A, Daroit J, Paula A, Correa F, Segalin J, Brandelli A. Production, one step purification, and characterization of a keratinolytic protease from *Serratia marcescens* P3. Process Biochemistry. 2012;**47**(12):2455-2462

[77] Korniłłowicz-K T, Bohacz J. Biodegradation of keratin waste: Theory and practical aspects. Waste Management. 2011;**31**:1689-1701

[78] Daroit DJ, Corrêa APF, Brandelli A. Keratinolytic potential of a novel *Bacillus* sp. P45 isolated from the Amazon basin fish *Piaractus mesopotamicus*. International Biodeterioration & Biodegradation. 2009;**63**:358-363

[79] Anitha TS, Palanivelu P. Purification and characterization of an extracellular keratinolytic protease from a new isolate of *Aspergillus parasiticus*. Protein Expression and Purification. 2013;**88**:214-220

[80] Deivasigamani B, Alagappan KM. Industrial application of soluble proteins from feather keratins. Journal of Environmental Biology. 2008;**29**(6):933-936

[81] Langeveld JPM, Wang J-J, de Wiel DFMV, Shih GC, Garssen J, Bossers A, Shih JCH. Enzymatic degradation of prion protein in brain stem from infected cattle and sheep. The Journal of Infectious Diseases. 2003;**188**:1782-1789

[82] Yoshioka M, Miwa T, Horii H, Takata M, Yokoyama T, Nishizawa K, Watanabe M, Shinagawa M, Murayama Y. Characterization of a proteolytic enzyme derived from a *Bacillus* strain that effectively degrades prion protein. Journal of Applied Microbiology. 2007;**102**:509-515

[83] Johnson KS, inventors; Ewa Beach HI, US. Degradation of beta amyloid proteins with keratinase. US Patent 2017/0042980 A1; 2017

[84] Mokrejs P, Svoboda P, Hrncirik J, Janacova D, VAsek V. Processing poultry feathers into keratin hydrolysate through alkaline-enzymatic hydrolysis. Waste Management & Research. 2011;**29**:260-267

[85] Lin X, Inglis GG, Yanke L, Cheng KJ. Selection and characterization of feather-degrading bacteria from canola meal compost. Journal of Industrial Microbiology & Biotechnology. 1999;**23**:149-153

[86] Gurav RJ, Jadhav JP. A novel source of biofertilizer from feather biomass for banana cultivation. Environmental Science and Pollution Research. 2013;**20**:4532-4539

[87] Hansen R, Keener HM, Marugg C, Dick WA, Hoitink HAJ. Composting of poultry manure. In: Science and Engineering of Composting: Design, Environmental, Microbiological and Utilization Aspects. Worthington, Ohio: Renaissance Publishers; 1993. pp. 131-153

[88] Narendra R. Non-food industrial applications of poultry feathers. Waste Management. 2015;**45**:91-107

[89] Tamreihao K, Devi LJD, Khunjamayum R, Mukherjee S, Ningthoujam D. Biofertilizing potential of feather hydrolysate produced by indigenous keratinolytic *Amycolatopsis sp.* MBRL 40 for rice cultivation under field conditions. Biocatalysis and Agricultural Biotechnology. 2017:317-320

[90] Choi J-M, Nelson PV. Developing a slow-release nitrogen fertilizer from organic sources: II. Using poultry feathers. Journal of the American Society for Horticultural Science. 1996;**121**:634-638

[91] Moore TJ. The cholesterol myth. The Atlantic. 1989;**264**:37

[92] Holick MF. Vitamin D: Importance in the prevention of cancers, type 1 diabetes, heart disease, and osteoporosis. The American Journal of Clinical Nutrition. 2004;**79**(3):362-371

[93] Ambrose CG, Clanton TO. Bioabsorbable implants: Review of clinical experience in orthopedic surgery. Annals of Biomedical Engineering. 2004;**32**(1):171-177

[94] Rouse JG, Dyke MEV. A review of keratin-based biomaterials for biomedical applications. Materials. 2010;**3**:999-1014

[95] Poole AJ, Church JS, Huson MG. Environmentally sustainable fibers from regenerated protein. Biomacromolecules. 2009;**10**:1-8

[96] Yin X-C, Li F-Y, He Y-F, Wang Y, Wang R-M. Study on effective extraction of chicken feather keratins and their films for controlling drug release. Biomaterials Science. 2013;**1**:528

[97] Xu H, Shi Z, Reddy N, Yang Y. Intrinsically water-stable keratin nanoparticles and their *in vivo* biodistribution for targeted delivery. Journal of Agricultural and Food Chemistry. 2014;**62**:9145-9150

[98] Edwards A, Jarvis D, Hopkins T, Pixley S, Bhattarai N. Poly(Ɛ-caprolactone)/keratin-based composite nanofibers for biomedical applications. Journal of Biomedial Materials Research Part B Applied Biomaterials. 2015;**103B**:21-30

[99] Boakye MAD, Rijal NP, Adhikari U, Bhattarai N. Fabrication and characterization of electrospun PCL-MgO-keratin-based composite nanofibers for biomedical applications. Materials. 2015;**8**:4080-4095

[100] Lee H, Noh K, Lee SC, Kwon IK, Han DW, Lee IS, Hwang YS. Human hair keratin and its-based biomaterials for biomedical applications. Tissue Engineering and Regenerative Medicine. 2014;**11**:255-265

[101] Hartrianti P, Ling L, Goh LMM, Ow KSA, Samsonraj RM, Sow WT, et al. Modulating mesenchymal stem cell behaviour using human hair keratin-coated surfaces. Stem Cells International. 2015:1-10

[102] Fakhfakh N, Gargouri M, Dahmen I, Sellami-Kamoun A, Feki AE, Nasri M. Improvement of antioxidant potential in rats consuming feathers protein hydrolysate obtained by fermentation of the keratinolytic bacterium, *Bacillus pumilus* A1. African Journal of Biotechnology. 2012;**11**:938-949

[103] Fontoura R, Daroit DJ, Correa APF, Meira SMM, Mosquera M, Brandelli A. Production of feather hydrolysates with antioxidant angiotensin-I converting enzyme and dipeptidyl peptidase-IV-inhibitory activities. New Biotechnology. 2014;**31**:506-513

[104] Coward-Kelly G, Agbogbo FK, Holtzapple MT. Lime treatment of shrimp head waste for the generation of highly digestible animal feed. Bioresource Technology. 2006;**97**:1515-1520

[105] Dalev P, Ivanov I, Liubomirova A. Enzymic modification of feather keratin hydrolysates with lysine aimed at increasing the biological value. Journal of the Science of Food and Agriculture. 1997;**73**:242-244

[106] Brandelli A, Sala L, Kalil SJ. Microbial enzymes for bioconversion of poultry waste into added-value products. Food Research International. 2015;**73**:3-12

[107] Weighmann H-D, Kamath YK, Ruetsch SB. Characterization of surface deposits on human hair fibers. Journal of the Society of Cosmetic Chemists. 1990;**41**:379-390

[108] Barba C, Mendez S, Roddick-Lanzilotta A, Kelly R, Parra J, Coderch L. Cosmetic effectiveness of topically applied hydrolysed keratin peptides and lipids derived from wool. Skin Research and Technology. 2008;**14**(2):243-248

[109] Villa AL, Aragao MR, Dos Santos EP, Mazotto AM, Zingali RB, de Souja EP, Vermelho AB. Feather keratin hydrolysates obtained from microbial keratinases: Effect on hair fibre. BMC Biotechnology. 2013;**1472-6750**(13):1-11

[110] Niinimaki A, Niinimaki M, Makinen-Kiljunen S, Hannuksela M. Contact urticaria from protein hydrolysates in hair conditioners. Allergy. 1998;**53**(11):1078-1082

[111] Bonser RH, Dawson C. The structural mechanical properties of down feathers and biomimicking natural insulation materials. Journal of Materials Science Letters. 1999; **18**(21):1769-1770

[112] Guan JP, Chen JQ. Flame retardancy finish with an organophosphorus retardant on silk fabrics. Fire and Materials. 2006;**30**:415-424

[113] Yang Y, Reddy N. Potential of using proteins and chicken feathers for cotton warp sizing. Cellulose. 2013;**20**:2163-2174

[114] Reddy N, Chen L, Zhang Y, Yang Y. Reducing environmental pollution of the textile industry using keratin as alternative sizing agent to poly (vinyl alcohol). Journal of Cleaner Production. 2014;**65**:561-567

[115] Brownstone S. Green Chemists use Rocket Science and Chicken Feathers to Create Cow-Less Leather. The Fast Company; 2013

[116] Enqi J, Narendra R, Zhifeng Z, Yiqi Y. Graft polymerization of native chicken feathers for thermoplastic applications. Journal of Agricultural and Food Chemistry. 2011; **59**:1729-1738

[117] Zhan M, Wool RP. Mechanical properties of chicken feather fibers. Polymer Composites. 2011;**32**:937-944

[118] Balaji S, Kumar R, Sripriya R, Rao U, Mandal A, Kakkar P, Reddy PN, Sehgal PK. Characterization of keratin-collagen 3D scaffold for biomedical applications. Polymers for Advanced Technologies. 2012;**23**:500-507

[119] Poole AJ, Church JS, Huson MG. Environmentally sustainable fibres from regenerated protein. Biomacromolecules. 2008;**10**(1):1-8

[120] Misra M, Kar P. Keratin fibers and structures for nanofiltration. Natural fibers, plastics and composites. 2003;**6**:83-93

[121] Misra M, Priyadarshan PKG. Keratin protein nano-fiber for removal of heavy metals and contaminants. Materials Research Society Symposium Proceedings. 2002;**702**:1-7

[122] Bálint B, Bagi Z, Tóth A, Rákhely G, Perei K, Kovács L. Utilization of keratin-containing biowaste to produce biohydrogen. Applied Microbiology and Biotechnology. 2005;**69**:404-410

Chapter 9

Keratin Waste: The Biodegradable Polymers

Tarun Kumar Kumawat, Anima Sharma,
Vishnu Sharma and Subhash Chandra

Additional information is available at the end of the chapter

http://dx.doi.org/10.5772/intechopen.79502

Abstract

Keratins are everywhere, from being the major components of household dust to common contaminants of laboratory protein analysis. Keratin is the major structural fibrous protein belonging to the large family of structural proteins to form hair, wool, feathers, nails, and horns of many kinds of animals and has a high concentration of cysteine, 7–20% of the total amino acid residues, that form inter- and intramolecular disulfide bonds. Keratin wastes are considered as the environmental pollutants and produced mostly from the poultry farms, slaughterhouses, and leather industries. Keratin wastes are dumped, buried, used for landfilling, or incinerated and all these actions increase the threats of environmental hazards, pollution, negatively influence the public health, and increase greenhouse gases concentration. Nature has provided planet Earth with a variety of beneficial organisms. Soil is considered as a well-known source for the growth of keratinophilic microflora (fungi and bacteria), which have the capability to degrade the keratin waste. The keratin-degradation ability of keratinophilic microflora has been credited with the production of the microbial keratinase enzyme and biodegradation takes place (enzymatic degradation). So, the keratin wastes are the biodegradable polymers. Keratinase is the industrially significant enzyme that offers bioconversion of keratin waste, utilization as animal feed supplements, and dehairing agents in tannery industries and textile industries.

Keywords: keratin, environmental pollutants, keratinophilic microflora, biodegradable polymers, keratinase

1. Introduction

Keratin wastes are considered as the environmental pollutants and generated mostly from the poultry farms, slaughterhouses and leather industries [1, 2]. The poultry farms, slaughterhouses, leather industries and wool industries are constantly producing a million tons of keratin waste [3]. Leather industries throw out extensive amount of waste products and considered as polluting industries with negative environmental impact [4]. The main producer of keratin waste includes the United State of America, China, India and Brazil which produces millions of tons of keratin containing protein [5, 6].

Keratin protein is the major component of the keratin waste [7] and belongs to the scleroprotein group [8]. Keratin protein is greatly resistant to the action of physical, chemical and biological agents [9]. The poultry feathers and other keratin-containing waste is dumped, land filled and incinerated throughout the world [10, 11]. These activities cause soil, water and air pollution. Discarded feather furthermore causes different human diseases including chlorosis and fowl cholera [12].

Very few microorganisms are capable to break down the keratin protein and utilize them as a source of nutrition [13]. Nature has provided earth with an assortment of beneficial organisms. Keratinophilic microflora (fungi and bacteria) is nature's gift and the biggest group of the organisms which have the capability to degrade the keratin waste [14, 15]. Biological degradation of keratin waste is more efficient than the physical and chemical degradation, yielding more useful by-product which can be utilized in commercial applications. In this scenario, biological keratin waste degradation has received the consideration from the scientific research community in recent days [16, 17].

2. Keratin protein

Keratin is an insoluble protein that forms the major component of the outer layer of the epidermis and helps to prevent the loss of body fluids [9, 14, 18]. Keratin word initially comes into view in the literature around 1850 to describe that keratin is made up from the hard tissues [19]. Keratin is the most complex proteins of epithelial cells of vertebrates [20, 21]. Keratin protein is a tough, fibrous and the third most abundant polymer in the environment after cellulose and chitin [22, 23].

According to the sulfur content, keratin proteins are divided into (a) soft keratins – skin and callus (b) hard keratins – feather, hair, hoof and [24–26]. This protein belongs to the scleroprotein group [27, 28]. The durability of keratins is a direct consequence of their complex architecture with extremely high molecular weight [29, 30]. Keratin protein is not easily degraded by pepsin, trypsin and papain because of disulfide bonds, hydrogen bonding, hydrophobic interactions [31–34].

2.1. Types of keratin

There are two types of keratin.

2.1.1. *Alpha keratin (α-keratin)*

Alpha keratin is found in the epithelia of all vertebrates [35]. The α-helix in alpha keratin constitutes the environmental problem due to their resistance to degradation from microbes [36, 37].

Alpha keratins in particular are remarkable for their strength, elasticity, toughness, insolubility and flexibility. Alpha (α) keratin has abundant quantities of hydrophobic amino acid, i.e. methionine, phenylalanine, valine, isoleucine and alanine [38]. According to the sulfur content, this protein is classified in hard and soft keratins [39].

2.1.2. Beta keratin (β-keratin)

Beta-keratin is structural protein and present in reptiles and birds [40]. Beta (β) keratin has high cysteine percentage and the cysteine readily forms disulfide bonds, which confer rigidity and provide enhanced resistance to degradation [41]. In a mature feather about 80–90% of β-keratin is present [42]. The molecular weight of individual keratin proteins is usually in the 10–14 kDa range [43, 44].

3. Major source of keratin protein

Keratin protein derives from living organism or from their body parts after death. The richest sources of keratin are feathers, wool, hair, hoof, scales and stratum corneum (**Figure 1**) [27]. Hair is the byproduct from tanneries during the haircut process [45]. Keratin protein is present in the human hair and offers flexibility, strength and durability to the hair in the form of different conformations [46, 47]. The bird's feather is made up of over 90% of keratin protein and produced as waste by poultry-processing industries [48].

Figure 1. Major sources of keratin protein (A) Bird's beak; (B) animal hair; (C) human nail; (D) horn; (E) human hair; (F) hoof; (G) nail; (H) chicken feather: the hosts for these sources include human, bird and animal.

The human hair is a natural filamentous biomaterial and chemically, approximate 80% keratin protein is present in human hair [49]. The accumulation of hair causes many environmental problems and considered as waste protein [50]. Feathers protect the birds from cold, rain, sun and injury [51]. The chicken feather is composed of about 90% of keratin [52], which is a fibrous and insoluble structural protein consisting of β-helical coils joined together by disulfide linkages [53]. This structural feature enables it to resist adverse environmental conditions and degrades by proteases [16]. Therefore, feathers are considered as a biological waste and cause serious environmental problems [54].

The human nail is an important organ of the human body and primarily composed of a highly cross-linked keratin network, a scleroprotein containing large amounts of sulfur (3.8%) with several disulfide linkages. This unique structure results in a highly effective permeability barrier [55]. The beak of the birds has an external shell of hard keratin which consists almost entirely of proteins [56]. Structurally, hoof keratin contains α-helical conformation with an admixture of β-sheet and possesses high thermal stability [57].

The horn is the tough animal tissue and has inflexible configuration due to the sulfur cross-linkages [58, 59]. Fundamental components of any horns are keratin, free amino acids, peptides, lipids, remain microelements: calcium, aluminum, chromium, copper, iron, manganese, and zinc [60]. Keratin protein in the animal horn is the tough-fiber, and its treatment is very difficult [61].

4. Impact of keratin waste on environmental pollution and human health

Industry has become an essential part of modern society, and waste production is an inevitable outcome of the developmental activities. Keratin wastes are produced in huge quantities from commercial poultry processing plants, leather industries, wool industry, textile industry, and slaughterhouses (**Figure 2**). These wastes may pose a potential hazard to the human health or the environment (soil, air, water) [1].

4.1. Keratin waste from the poultry industry

Feathers from the chicken generated in large quantity as a waste by-product of the poultry processing plant. Worldwide, around 8.5 billion tons of poultry feather is generated annually, of which India's contribution alone is 350 million tons. Accumulation of chicken feathers will lead to environmental contamination [12, 62, 63]. Chicken feather causes environmental pollution as well as adversely affects the people's life living in nearby localities [64].

4.2. Keratin waste from slaughterhouses

Keratin waste is generated from the meat industry (slaughterhouses) in the form of chicken feathers, beaks, mixture of bones, organs and hard tissues in very large quantity. Keratinous wastes are degraded very slowly in nature, and considered as hazardous wastes according to EU directives [65]. The contaminated waste water generated from such industries caused the problems of acidification of soils, eutrophication and decreased species diversity. The conventional methods employed for the disposal of keratin waste are not only costly, but also very

Figure 2. Major keratin waste producing industries.

difficult. Decomposition methods like incineration are employed [20], but these procedures are environment-polluting and pose risk to the environment [66].

4.3. Keratin waste from leather industry

Leather industries are the most polluting industries globally. The leather processing is responsible for unfavorable impact on the environment [67, 68]. Keratin wastes generated from leather industries in very large amounts include both solid and liquid waste, which is mostly of animal origin [69]. A considerable amount of keratin protein waste such as hair, horns and hoofs are thrown away by leather industries [20]. Tannery industries discharge the wastes and causing serious health problems as well as pollute the air, soil, and water [70].

4.4. Keratin waste from barber shops

Barber and hair stylist shops are also the most important keratin pollution sources. Human hair is considered as environment pollutant and found as the municipal waste in the world [71]. In the city area, it often accumulates in large amounts as solid waste and chokes the drainage systems. In rural areas, hair is thrown away in nature where it slowly decomposes over several years. Open dumps of hair generate hair-dust which causes discomfort to people residing in these areas and, if inhaled in large amounts, can result in several respiratory problems [50].

5. Traditional disposal strategies of keratin waste and their disadvantages

Each year, approximately 24 billion chickens are killed across the world and huge amount of poultry feathers produces globally [72, 73], in addition to the accumulation of human hair in waste treatment facilities worldwide [51]. Keratin solid waste generated from meat, poultry processing, fish industries, wool industries considered as harmful environmental pollutants [74, 75].

Due to pathogenic microbes on the keratin waste, efficient and immediate treatment of keratin waste has become necessary [76]. The tremendous volume of keratin waste creates a serious solid waste problem in many countries [77]. The keratin waste is linked with the evolution of odors and pathogens into the soil and water [78]. Disposal of keratin waste is quite challenging [79].

Considering the huge quantity generated, there are four methods for dealing with keratin waste: incineration, landfilling, composting, and mechanical grinding (**Figure 3**) [20].

5.1. Incineration

Incineration involves combustion of keratin waste and destroying potential infectious agents [80]. Incineration plant's temperatures are above 850°C and mostly waste is converted to CO_2 and water [81]. Due to the requirement of high temperature, the operating costs are not only expensive but also difficult to maintain [6]. Incineration leads to the release of pollutants into the atmosphere, causing foul odors and contribute to harmful runoff, which negatively impacts the surrounding and downstream areas including livestock and nearby ecosystems [50].

5.2. Landfilling

The traditional method for disposal of keratin wastes is land filling [20, 79]. Historically, landfills have been the most popular methods of organized waste disposal and continue to remain in several places around the world [82]. The improper disposal of keratin wastes by landfilling contributes to environmental damage and transmission of diseases [83]. Land filling also poses problems like landfill leachate and greenhouse gases [84]. Leachate increases the nitrogen concentration in surrounding areas, leading to algal blooms and harming the ecosystem [50]. So, landfilling is the less expensive way for discarding of keratins waste, but it is not an efficient method.

5.3. Composting

Composting is the additional economical method for recycling feather waste. Ninety percent of the feather-weight consists of crude keratin protein, and also contain 15% N [85, 86]. Composting is an aerobic biological process degrading organic material of poultry, slaughterhouse wastes, manure, and litter. This process reduces the pathogens, and compost product can be used as the soil fertilizer [87–89].

5.4. Mechanical grinding

The method to dispose of keratin waste is mechanically breaking it down into useful products. In this process, the poultry feathers are hydrolyzing under heat and pressure and then grinding and drying. The dried waste ground into a powder and later processed into useful products [90]. The ground powder can be used as a nitrogen source for animal feed (mostly ruminants) or as an organic soil enhancer [91]. There are certain disadvantages of the mechanical grinding method. Extremely high temperature and grinding result in the loss of several valuable amino acids [69, 92].

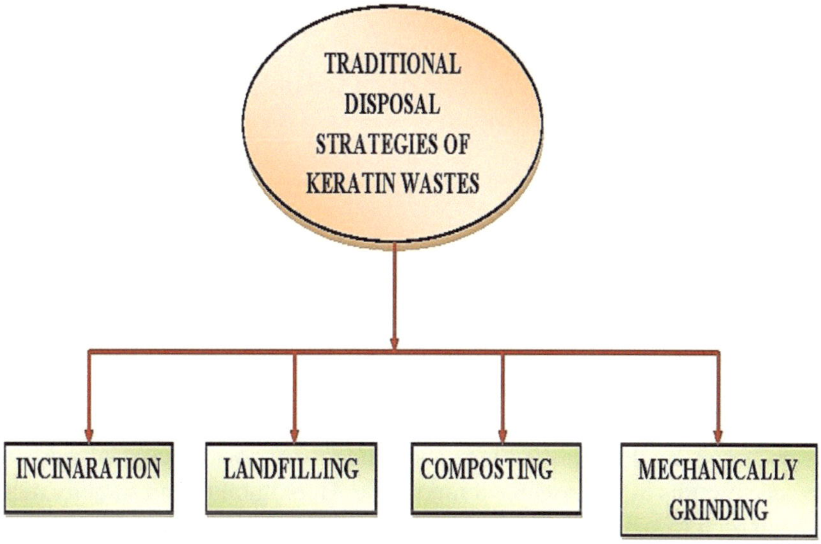

Figure 3. Traditional disposal methods of chicken feather waste.

Disposal of keratin waste from incineration, landfilling, composting, and mechanical grinding is restricted because of enormous production of harmful gases and poses the risk to the environment [66]. Considering the disadvantages of the above all methods, the management of keratin waste using microorganisms appears to be a viable option and it is therefore, attracting scientists for research in this field.

6. Techniques for hydrolysis of keratin waste

The management of keratin waste generated in the poultry industries, leather industries, and slaughterhouses is a major concern of many nations across the globe [93, 94]. Numbers of methods, including hydrothermal, chemical, enzymatic or biological treatments have been therefore investigated in the past few years to improve the digestibility of feathers (**Figure 4**) [20, 95, 96].

6.1. Hydrothermal method

The hydrothermal process usually employs high temperature (80–140°C) and high steam pressure (10–15 psi) with the addition of acids or bases for the degradation of keratin wastes [97, 98]. This method consumes the high quantity of energy and addition of acids (HCl) or bases (NaOH), which break peptide bonds of keratin [99, 100]. Hydrothermal hydrolysis of degradation also required a longer time (16 hours) for feather degradation [20].

Keratin protein is not degraded by trypsin, pepsin, and papain in its native state, because of multiple disulfide bonds [4]. Keratin waste is disposed of through thermal processing according to health regulations. The ash product that is obtained from this process is rich in macronutrients as well as micronutrients. These components have high fertilizing value [97]. The recent processes of hydrothermal treatment are costly as well as destroy amino acids and contain non-nutritive amino acids such as lanthionine and lysinoalanine [70, 101, 102].

6.2. Chemical method

The chemical hydrolysis process of keratin wastes is based on the chemicals (acid, base, catalyst). Chemical hydrolysis requires more aggressive conditions of the reaction (high temperature and pressure) and carries a greater risk to the environment [103]. The chemical hydrolysis reaction is slower and highly efficient, but causes the loss of some amino acids, e.g. tryptophan [76]. The chemical methods require more time, chemicals and energy with expensive industrial equipment for processing. The product has low nutritional value because it contains small amounts of the essential amino acids. The solubility and stability of the hydrolysates depend on the degree of protein degradation [99].

The chemical hydrolysis process increases the emission of certain gases like CO, SO_2 into the environment and causes respiratory diseases, cardiovascular diseases, and cancer, among other illnesses [16]. Hence, there is an urgent need to develop biotechnological and eco-friendly alternatives for recycling of keratin waste.

6.3. Biological method

Considering the potent polluting implications and thermo-energetic cost of the above approaches for the treatment of keratin waste, microbial degradation/ biological method is an alternative, cost-effective and ecologically safe method [104, 105]. Keratinase enzymes produced by microorganisms are the possible alternative to convert keratin waste into the nutrient-rich animal feed [106, 107]. Very few microorganisms utilize keratin by enzymatic digestion as a source of nutrient substrate for growth. These microorganisms are called keratinophilic microflora.

Keratinophilic microflora represents a significant component of soil and an important group of fungi, bacteria and insects that degrade the highly stable animal proteins on earth due to the release of keratinases [14, 34, 108, 109]. Microbial keratinase is a proteolytic enzyme that

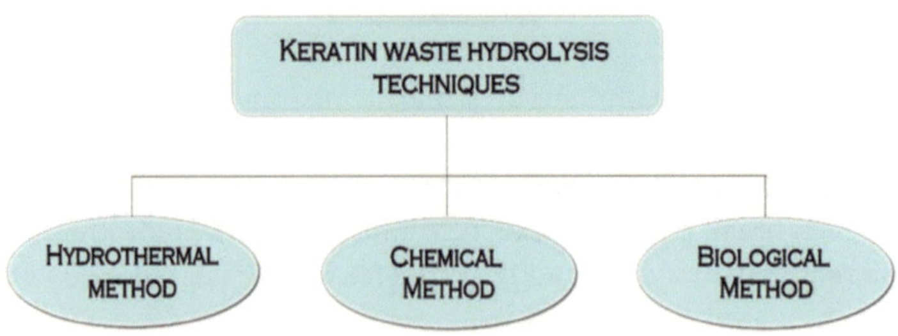

Figure 4. Keratin waste hydrolysis techniques.

posses the capability to degrade the insoluble keratin protein [110–112]. The enzymatic ability of keratin degrading microorganism to decompose keratin has long been interpreted as a key innovation [113]. Keratinase enzymes have molecular weights ranging from 18 to 240 kDa [114, 115].

Biological degradation of keratin waste is more efficient than hydrothermal and chemical degradation, resulting in more useful and toxin-free product. Thus, on employing this ability of keratinophilic microorganisms on an industrial scale, the environmental impacts of incineration and landfilling can be reduced to a great extent.

Microbial keratin degradation follows the sequence of adhesion, colonization, amplification of keratinase pursued by the breakdown and deprivation of the substrate [116]. In the process of microbial keratin degradation, microorganisms' preliminary consumes the lipids (non-keratinous elements) and then begin to degrade keratin [117]. Keratin degradation comprises two major actions, i.e., sulfitolysis (breakdown of disulfide bonds) and proteolysis (proteolytic attack) by keratinolytic proteases (keratinases) based on the complexity nature of keratin [16, 106, 109]. Sulfitolysis is the main process of keratinolysis [118]. In this process, microorganisms discharge sulfide, which is accountable for the breakdown of keratin's disulfide bonds [14]. In proteolysis, bacteria and fungi are able to degrade keratinous substrates proficiently due to their ability to secrete extracellular keratinase enzyme into the medium [22, 105].

Keratinophilic microbes attack keratin substrates in or on soil; therefore, biodegradation takes place [27, 119]. Several microbial strains could be valuable as they possess very significant degradation ability [94]. The keratin-degrading fungi are an environmentally important group of fungi and considered as soil saprophytes [120, 121]. The soil is rich in the keratin protein so the keratinophilic fungi easily occur and grow [122, 123]. Keratin degrading fungi colonize keratin waste and degrade them into low molecular weight [124, 125]. Most of the keratinophilic fungi belong to the families of Arthodermataceae and Onygenaceae in Ascomycetes. The keratinophilic species belong to the genera *Chrysosporium, Microsporum, Trichophyton, Aspergillus, Fusarium,* and *Uncinocarpus* [126]. These fungi are active producers of extracellular keratinase, they can be used in bioremediation of such waste and waste contaminated sites [14, 127].

A number of bacterial strains are capable of degrading keratins have been reported. Bacteria can grow faster than fungal species and therefore have potential in industrial applications. The degradation of keratin is predominantly confined by *Bacillus, Microbacterium, Lysobacter, Nesternokia* and *Kocuria* (Gram-positive bacteria) and *Vibrio,* and *Xanthomonas* and *Chryseobacterium* (Gram-negative bacteria) [128, 129]. The maximum feather-degrading abilities are observed mostly in the strains of *Bacillus licheniformis* [31, 130] and less frequently in populations of *Bacillus pumilus, Bacillus cereus* and *Bacillus subtilis* [131]. Keratin-degrading bacteria are *Burkholderia, Chryseobacterium, Pseudomonas, Microbacterium* sp., *Vibrio, Flavobacterium,* and *Thermoanaerobacter* [132, 133].

Studies on keratinophilic fungi started in 1952 with the discovery of hair baiting technique. This technique facilitated researchers for isolation of fungi from soil throughout the world [134].

Otcenasek [135] reported worldwide distribution of keratinophilic mycobiota in soil. Most keratin degrading fungi belong to Arthodermataceae and Onygenaceae families of the order Onygenales in Ascomycetes. The growth of fungi on temperature ranging from 15–35°C and some require a range of high temperature for optimum growth [29, 136]. Fungi grow at pH neutral to the weak acidic environment, with the highest production mycelial. Optimum pH 5.0–8.0 is suitable for conidial production and sporulation in liquid media [137, 138]. The screening of keratinolytic activity of fungi was tested through chicken feather degradation in Basal Salt Medium (BSM) [139].

Similar to isolates of fungi, lists of bacterial strains capable of degrading keratins have been reported [140]. Williams [141] isolated feather degrading Gram variable, endospore forming, motile, rod shaped bacterium and identified as *Bacillus licheniformis* PWD-1. This isolate demonstrated facultative growth at thermophilic temperatures with optimum at 45–50°C and pH 7.5. Deivasigamani and Alagappan [9] isolated keratinolytic *Bacillus* sp. from slaughterhouse and poultry farm and observed maximum keratinase activity (122.5 KU/ml) at pH 8.0. Cao [142] isolated a father degrading bacterium (*Stenotrophomonas maltophilia*) from decomposing poultry feathers, which showed the highest feather degrading activity at 40°C and pH 7.5–8.0. The keratin degrading microbes are widespread among the soil microbial population. These microbes have the ability to colonize and breakdown the complex keratinous waste.

The keratinophilic microorganisms effectively degrade the keratin waste and recycle them into valuable products [143]. The possible use of keratinase is in various applications such as in the poultry industries, waste bioconversion, leather industries, pharmaceutical industries, textile processing, detergent formulation, animal feed and fertilizers [144–146].

7. Conclusions

The keratinophilic microflora degrades the various keratinous waste effectively and showed the keratinolytic activity. The keratinous waste degradation by biological way is not only economical but also a possible process for better management of keratinous wastes. Keratin degrading microorganisms could be used for biotechnological application in recycling of poultry waste for environmental protection (production of nitrogenous fertilizer and animal feed) and its fermentation broth could be useful in leather industry and textile industry, etc.

Acknowledgements

We thank the Director, School of Sciences for encouragement, and the Head, Department of Biotechnology, JECRC University, Jaipur, for providing the laboratory facilities.

Author details

Tarun Kumar Kumawat[1], Anima Sharma[2]*, Vishnu Sharma[3] and Subhash Chandra[4]

*Address all correspondence to: sharmaanima6@gmail.com

1 Therachem Research Medilab (India) Pvt. Ltd., Jaipur, Rajasthan, India

2 Department of Botany, Maharshi Dayanand Saraswati University, Ajmer, Rajasthan, India

3 Department of Botany, Mehta PG College, Jaipur, Rajasthan, India

4 Department of Zoology, Maharshi Dayanand Saraswati University, Ajmer, Rajasthan, India

References

[1] Saber WIA, El-Metwally MM, El-Hersh MS. Keratinase production and biodegradation of some keratinous wastes by Alternaria tenuissima and Aspergillus nidulans. Research Journal of Microbiology. 2010;**5**(1):21-35

[2] Darah I, Nur-Diyana A, Nurul-Husna S, Jain K, Sheh-Hong L. Microsporum fulvum IBRL SD3: As novel isolate for chicken feathers degradation. Applied Biochemistry and Biotechnology. 2013;**171**(7):1900-1910

[3] Mokrejs P, Krejci O, Svoboda P, Vasek V. Modeling technological conditions for break-down of waste sheep wool. Rasayan Journal of Chemistry. 2011;**4**(4):728-735

[4] Gousterova A, Braikova D, Goshev I, Christov P, Tishinov K, Vasileva-Tonkova E, Haertle T, Nedkov P. Degradation of keratin and collagen containing wastes by newly isolated thermoactinomycetes or by alkaline hydrolysis. Letters in Applied Microbiology. 2005;**40**(5):335-340

[5] Sousa M, Souza O, Maciel M, Cruz R, Rego MG, Magalhaes O, Pessoa-Junior A, Porto A, Souza-Motta C. Keratinolytic potential of fungi isolated from soil preserved at the Micoteca URM. European Journal of Biotechnology and Bioscience. 2015;**3**(5):10-15

[6] Sharma S, Gupta A. Sustainable management of keratin waste biomass: Applications and future perspectives. Brazilian Archives of Biology and Technology. 2016;**59**:e16150684

[7] Godheja J, Shekhar SK. Biodegradation of keratin from chicken feathers by fungal species as a means of sustainable development. Journal of Bioremediation & Biodegradation. 2014;**5**(5):232

[8] Isaac GS, Abu-Tahon MA. Dehairing capability of alkaline keratinase produced by new isolated Cochliobolus hawaiiensis AUMC 8606 grown on chicken feather. Romanian Biotechnological Letters. 2016;**22**(6):12147-12154

[9] Deivasigamani B, Alagappan KM. Industrial application of keratinase and soluble proteins from feather keratins. Journal of Environmental Biology. 2008;**29**(6):933-936

[10] Agrahari S, Wadhwa N. Degradation of chicken feather a poultry waste product by keratinolytic bacteria isolated from dumping site at Ghazipur poultry processing plant. International Journal of Poultry Science. 2010;**9**(5):482-489

[11] Manirujjaman M, Amin R, Nahid AA, Alam MS. Isolation and characterization of feather degrading bacteria from poultry waste. African Journal of Bacteriology Research. 2016;**8**(3):14-21

[12] Williams CM, Lee CG, Garlich JD, Shih JCH. Evaluation of a bacterial feather fermentation product, feather-lysate, as a feed protein. Poultry Science. 1991;**70**(1):85-94

[13] Kunert J. Physiology of keratinophilic fungi. In: Kushwaha RKS, Guarro J, editors. Biology of Dermatophytes and Other Keratinophilic Fungi. Bilbao: Revista Iberoamericana de Micologia; 2000. pp. 77-85

[14] Sharma R, Rajak RC. Keratinophilic fungi: Nature's keratin degrading machines their isolation, identification and ecological role. Resonance. 2003;**8**(9):28-40

[15] Tridico SR, Koch S, Michaud A, Thomson G, Kirkbride KP, Bunce M. Interpreting biological degradative processes acting on mammalian hair in the living and the dead: Which ones are taphonomic? Proceedings of the Royal Society B. 2014;**281**:20141755

[16] Gupta R, Ramnani P. Microbial keratinases and their prospective applications: An overview. Applied Microbiology and Biotechnology. 2006;**70**(1):21-33

[17] Brandelli A. Bacterial Keratinases: Useful enzymes for bioprocessing agroindustrial waste and beyond. Food and Bioprocess Technology. 2008;**1**(2):105-116

[18] Sharma R, Swati. Effect of keratin substrates on the growth of keratinophilic fungi. Journal of Academia and Industrial Research. 2012;**1**(4):170-172

[19] Rouse JG, Dyke MEV. A review of keratin-based biomaterials for biomedical applications. Materials. 2010;**3**(2):999-1014

[20] Onifade A, Al-Sane NA, Al-Musalism AA, Al-Zarban S. A review: Potentials for biotechnological applications of keratin-degrading microorganisms and their enzymes for nutritional improvement of feathers and other keratins as livestock feed resources. Bioresource Technology. 1998;**66**(1):1-11

[21] Sapna R, Yamini V. Study of keratin degradation by some potential bacterial isolates from soil. Journal of Soil Science. 2011;**1**(1):01-03

[22] Lange L, Huang Y, Busk PK. Microbial decomposition of keratin in nature-a new hypothesis of industrial relevance. Applied Microbiology and Biotechnology. 2016;**100**(5): 2083-2096

[23] Kumawat TK, Sharma A, Bhadauria S. Chrysosporium queenslandicum: A potent keratinophilic fungus for keratinous waste degradation. International Journal of Recycling of Organic Waste in Agriculture. 2017;**6**(2):143-148

[24] Esawy MA. Isolation and partial characterization of extracellular keratinase from a novel mesophilic Streptomyces albus AZA. Research Journal of Agriculture and Biological Sciences. 2007;**3**(6):808-817

[25] Tork S, Aly MM, Nawar L. Biochemical and molecular characterization of a new local keratinase producing Pseudomomanas sp., MS21. Asian Journal of Biotechnology. 2010;**2**(1):1-13

[26] Saibabu V, Niyonzima FN, More SS. Isolation, partial purification and characterization of keratinasefrom *Bacillus megaterium*. International Research Journal of Biological Sciences. 2013;**2**(2):13-20

[27] Kim JD. Purification and characterization of a keratinase from a feather-degrading fungus, Aspergillus flavus strain K-03'. Mycobiology. 2007;**35**(4):219-225

[28] Sivakumar T, Balamurugan P, Ramasubramanian V. Characterization and applications of keratinase enzyme by Bacillus thuringiensis TS2. International Journal of Future Biotechnology. 2013;**2**(1):1-8

[29] Sharma M, Sharma M, Rao VM. In vitro biodegradation of keratin by dermatophytes and some soil keratinophiles. African Journal of Biochemistry Research. 2011;**5**(1):1-6

[30] Sowjanya NC, Chary CM. Degradation of few avean feathers by *Microsporum gypseum*. Journal of Phytology. 2012;**4**(4):21-23

[31] Lin X, Lee CG, Casale ES, Shih JCH. Purification and characterization of a keratinase from a feather-degrading *Bacillus licheniformis* strain. Applied and Environmental Microbiology. 1992;**58**(10):3271-3275

[32] Tapia DMT, Contier J. Production and partial characterization of keratinase produced by a microorganism isolated from poultry processing plant wastewater. African Journal of Biotechnology. 2008;**7**(3):296-300

[33] Lin HH, Yin LJ. Feather meal and rice husk enhanced keratinases production by *Bacillus licheniformis* YJ4 and characters of produced keratinases. Journal of Marine Science and Technology. 2010;**18**(3):458-465

[34] Sinoy TES, Bhausaheb CP, Rajendra PP. Isolation and identification of feather degradable microorganism. VSRD Technical & Non-Technical Journal. 2011;**2**(3):128-136

[35] Vandebergh W, Bossuyt F. Radiation and functional diversification of alpha keratins during early vertebrate evolution. Molecular Biology and Evolution. 2012;**29**(3):995-1004

[36] Kannahi M, Ancy RJ. Keratin degradation and enzyme producing ability of *Aspergillus flavus* and *Fusarium solani* from soil. Journal of Chemical and Pharmaceutical Research. 2012;**4**(6):3245-3248

[37] Balakumar S, Mahesh N, Arunkumar M, Sivakumar R, Hemambujavalli V. Optimization of keratinase production by keratinolytic organisms under submerged fermentation. International Journal of PharmTech Research. 2013;**5**(3):1294-1300

[38] Lehninger AB, Nelson DL, Cox MM. Principles of Biochemistry. 2nd ed. New York: Worth Publishers; 1993

[39] Nickerson WJ. Biology of Pathogenic Fungi. University of Michigan: Chronica Botanica Co; 1947

[40] Greenwold MJ, Sawyer RH. Molecular evolution and expression of archosaurian β-keratins: Diversification and expansion of archosaurian β-keratins and the origin of feather β-keratins. Journal of Experimental Zoology Part B: Molecular and Developmental Evolution. 2013;**320**(6):393-405

[41] Strasser B, Mlitz V, Hermann M, Tschachler E, Eckhart L. Convergent evolution of cysteine-rich proteins in feathers and hair. BMC Evolutionary Biology. 2015;**15**:82

[42] Cai C, Zheng X. Medium optimization for keratinase production in hair substrate by a new Bacillus subtilis KD-N2 using response surface methodology. Journal of Industrial Microbiology and Biotechnology. 2009;**36**(7):875-883

[43] Woodin AM. Structure and composition of soluble feather keratin. Biochemical Journal. 1956;**63**(4):576-581

[44] Shames RB, Knapp LW, Carver WE, Washington LD, Sawyer RH. Keratinization of the outer surface of the avian scutate scale: Interrelationship of alpha and beta keratin filaments in a cornifying tissue. Cell and Tissue Research. 1989;**257**(1):85-92

[45] Cranston RW, Davis MH, Scroggie JG. Development of the sirolime unhairing process. Journal of the American Leather Chemists Association. 1986;**81**(11):347-355

[46] Wagner RCC, Joekes I. Hair protein removal by sodium dodecyl sulfate. Colloids and Surfaces B: Biointerfaces. 2007;**41**(1):7-14

[47] Velasco MVR, Dias TCD, De Freitas AZ, Vieira ND, Pinto CASD, Kaneko TM, Baby AR. Hair fiber characteristics and methods to evaluate hair physical and mechanical properties. Brazilian Journal of Pharmaceutical Sciences. 2009;**45**(1):153-162

[48] Nagal S, Jain PC. Feather degradation by strains of *Bacillus* isolated from decomposing feathers. Brazilian Journal of Microbiology. 2010;**41**(1):196-200

[49] Kaplin IJ, Schwan A, Zahn H. Effects of cosmetic treatments on the ultrastructure of hair. Cosmetics & Toiletries. 1982;**97**(8):22-25

[50] Gupta A. Human hair "waste" and its utilization: Gaps and possibilities. Journal of Waste Management. 2014. Article ID: 498018. 17 p. DOI: 10.1155/2014/498018

[51] Adetola SO, Yekini AA, Olayiwola BS. Investigation into physical and mechanical properties of few selected chicken feathers commonly found in Nigeria. IOSR Journal of Mechanical and Civil Engineering. 2014;**11**(3):45-50

[52] Bishmi A, Thatheyus J, Ramya D. Biodegradation of poultry feathers using a novel bacterial isolate *Pseudomonas aeruginosa*. International Journal of Research Studies in Microbiology and Biotechnology. 2015;**1**(1):25-30

[53] DaGioppo NM, Moreira FG, Costa AM, Alexandrino AM, De Souza CG, Peralta RM. Influence of the carbon and nitrogen sources on keratinase production by *Myrothecium verrucaria* in submerged and solid state cultures. Journal of Industrial Microbiology and Biotechnology. 2009;**36**(5):705-711

[54] Chaturvedi V, Verma P. Metabolism of chicken feathers and concomitant electricity generation by *Pseudomonas aeruginosa* by employing microbial fuel cell (MFC). Journal of Waste Management. 2014. Article ID: 928618. 9 p. DOI: 10.1155/2014/928618

[55] Gupchup GV, Zatz JL. Structural characteristics and permeability properties of the human nail: A review. Journal of the Society of Cosmetic Chemists. 1999;**50**(6):363-385

[56] Frenkel MJ, Gillespie JM. 'The proteins of the keratin component of bird's beaks. Australian Journal of Biological Sciences. 1976;**29**(5-6):467-479

[57] Kakkar P, Madhan B, Shanmugam G. Extraction and characterization of keratin from bovine hoof: A potential material for biomedical applications. Springer Plus. 2014;**3**(1):596

[58] Mills JS, White R. The Organic Chemistry of Museum Objects. London, UK: Butterworth-Heineman; 1994

[59] Mikulikova K, Romanov O, Miksik I, Eckhardt A, Pataridis S, Sedlakova P. Study of saiga horn using high-performance liquid chromatography with mass spectrometry. The Scientific World Journal. 2012. Article ID: 759604. 8 p. DOI: 10.1100/2012/759604

[60] Romanov OE. The comparative analysis of horn properties and their composition. Scientific Research of Caucasus. 2005:91-95

[61] Kida K, Morimura S, Noda J, Nishida Y, Imai T, Otagiri M. Enzymatic hydrolysis of the horn and hoof of cow and buffalo. Journal of Fermentation and Bioengineering. 1995;**80**(5):478-484

[62] Mohamedin AH. Isolation identification and some cultural conditions of a protease producing thermophilic Streptomyces strain grown on chicken feather as a substrate. International Biodeterioration & Biodegradation. 1999;**43**(1-2):13-21

[63] Huda S, Yang Y. Composites from ground chicken quill and polypropylene. Composites Science and Technology. 2008;**68**(3-4):790-798

[64] Gerber P, Opio C, Steinfeld H. Poultry Production and the Environment – A Review. Viale delle Terme di Caracalla. Room, Italy: Animal Production and Health Division, Food and Agriculture Organization of the United Nations; 2007. p. 153

[65] Deydier E, Guilet R, Sarda S, Sharrock P. Physical and chemical characterisation of crude meat and bone meal combustion residue: Waste or raw material? Journal of Hazardous Materials. 2005;**121**(1-3):141-148

[66] Verheyen L, Wiersema D, Hulshoff-Pol L, Brandjes P, Westra P, Bos J, Jansen J. Management of Waste from Animal Product Processing. In: Livestock and Environment, Finding a Balance. Wageningen, The Netherlands: Internal Agriculture Centre; 1996

[67] Kanagaraj J, Velappan KC, Chandra BNK, Sadulla S. Solid wastes generation in the leather industry and its utilization for cleaner environment – A review. Journal of Scientific and Industrial Research. 2006;**65**(7):541-548

[68] Mushahary I, Mirunalini V. Waste management in leather industry – Environmental and health effects and suggestions to use in construction purposes. International Journal of Civil Engineering & Technology. 2017;**8**(4):1394-1401

[69] Wang X, Parsons CM. Effect of processing systems on protein quality of feather meal and hair meals. Poultry Science. 1997;**76**(3):491-496

[70] Syed M, Saleem T, Rehman S, Iqbal MA, Javed F, Khan MB, Sadiq K. Effects of leather industry on health and recommendations for improving the situation in Pakistan. Archives of Environmental and Occupational Health. 2010;**65**(3):163-172

[71] Kumar S, Bhattacharyya JK, Vaidya AN, Chakrabarti T, Devotta S, Akolkar AB. Assessment of the status of municipal solid waste management in metro cities, state capitals, class I cities, and class II towns in India: An insight. Waste Management. 2009;**29**(2):883-895

[72] Jagadeeshgouda KB, Reddy P, Ishwaraprasad K. Experimental study of behaviour of poultry feather fiber – A reinforcing material for composites. International Journal of Research in Engineering and Technology. 2014;**3**(2):362-371

[73] Ashwathanarayana R, Shashidhara TJ, Naika R. Determination of keratinolytic potential of keratinolytic fungi on poultry feather samples isolated from selected poultry sites around Shivamogga, Karnataka, India. International Journal of Science and Research Methodology. 2015;**2**(2):6-13

[74] Jayathilakan K, Sultana K, Radhakrishna K, Bawa AS. Utilization of byproducts and waste materials from meat, poultry and fish processing industries: A review. Journal of Food Science and Technology. 2012;**49**(3):278-293

[75] Kumawat TK, Sharma A, Bhadauria S. Biodegradation of keratinous waste substrates by Arthroderma multifidum. Asian Journal of Applied Sciences. 2016;**9**(6):106-112

[76] Sinkiewicz I, Sliwinska A, Staroszczyk H, Kolodziejska I. Alternative methods of preparation of soluble keratin from chicken feathers. Waste and Biomass Valorization. 2017;**8**(4):1043-1048

[77] McGovern V. Recycling poultry feather: More bang for the cluck. Environmental Health Perspective. 2000;**108**(8):336-339

[78] Tufaner F, Avsar Y. Effects of co-substrate on biogas production from cattle manure: A review. International journal of Environmental Science and Technology. 2016;**13**(9): 2303-2312

[79] Mehta RS, Jholapara RJ, Sawant CS. Isolation of a novel feather-degrading bacterium and optimization of its cultural conditions for enzyme production. International Journal of Pharmacy and Pharmaceutical Sciences. 2014;**6**(1):194-201

[80] Ritter WF, Chinside AEM. Impact of dead bird disposal pits on groundwater quality on the Delmarva Peninsula. Bioresource Technology. 1995;**53**(2):105-111

[81] Dube R, Nandan V, Dua S. Waste incineration for urban India: Valuable contribution to sustainable MSWM or inappropriate high-tech solution affecting livelihoods and public health? International Journal of Environmental Technology and Management. 2014;**17**(2/3/4):199-214

[82] Remigios MV. An overview of the management practices at solid waste disposal sites in African cities and towns. Journal of Sustainable Development in Africa. 2010;**12**(7):233-239

[83] Tronina P, Bubel F. Production of organic fertiliser from poultry feather wastes excluding the composting process. Polish Journal of Chemical Technology. 2008;**10**(2):33-36

[84] Vuppu S, Sinha R, Gupta A, Goyal R. An attempt and a brief research study to produce mosquitocidal toxin using Bacillus spp. (VITRARS), isolated from different soil samples (Vellore and Chittoor), by degradation of chicken feather waste. Research Journal of Pharmaceutical, Biological and Chemical Sciences. 2012;**3**(4):40-48

[85] Ichida JM, Krizova L, LeFevre CA, Keener HM, Elwell DL, Burtt EH. Bacterial inoculum enhances keratin degradation and biofilm formation in poultry compost. Journal of Microbiological Methods. 2001;**47**(2):199-208

[86] Tiquia SM. Evaluation of organic matter and nutrient composition of partially decomposed and composted spent pig litter. Environmental Technology. 2003;**24**(1):97-107

[87] Davalos JZ, Roux MV, Jimenez P. Evaluation of poultry litter as a feasible fuel. Thermochimica Acta. 2002;**394**(1-2):261-266

[88] Thyagarajan D, Barathi M, Sakthivadivu R. Scope of poultry waste utilization. IOSR Journal of Agriculture and Veterinary Science. 2013;**6**(5):29-35

[89] Liu J, Luo Q, Huang Q. Removal of 17 β-estradiol from poultry litter via solid state cultivation of lignolytic fungi. Journal of Cleaner Production. 2016;**139**:1400-1407

[90] Jaouadi NZ, Rekik H, Badis A, Trabelsi S, Belhoul M, Yahiaoui AB, Aicha HB, Toumi A, Bejar S, Jaouadi B. Biochemical and molecular characterization of a serine keratinase from Brevibacillus brevis US575 with promising keratin-biodegradation and hide-dehairing activities. PLoS One. 2013;**8**(10):e76722

[91] Hadas A, Kautsky L. Feather meal, a semi-slow-release nitrogen fertilizer for organic farming. Fertilizer Research. 1994;**38**(2):165-170

[92] Latshaw JD, Musharaf N, Retrum R. Processing of feather meal to maximize its nutritional value for poultry. Animal Feed Science and Technology. 1994;**47**(3-4):179-188

[93] Schmidt WF, Barone JR. New uses for chicken feathers keratin fiber. In: National Poultry Waste Management Symposium, At Memphis TN. 2004. pp. 99-101

[94] Lateef A, Oloke JK, Gueguim KEB, Sobowale BO, Ajao SO, Bello BY. Keratinolytic activities of a new feather-degrading isolate of *Bacillus cereus* LAU 08 isolated from Nigerian soil. International Biodeterioration & Biodegradation. 2010;**64**(2):162-165

[95] Papadopoulos MC. Effect of processing on high-protein feedstuffs: A review. Biological Waste. 1989;**29**(2):123-138

[96] Staron P, Kowalski Z, Staron A, Banach M. Thermal treatment of waste from the meat industry in high scale rotary kiln. International journal of Environmental Science and Technology. 2017;**14**(6):1157-1168

[97] Karthikeyan R, Balaji S, Sehgal PK. Industrial applications of keratins – A review. Journal of Scientific and Industrial Research. 2007;**66**(9):710-715

[98] Onuoha SC, Chukwura EI. Effect of temperature and pH on bacterial degradation of chicken feather waste (CFW). International Journal of Science and Nature. 2011;**2**(3): 538-544

[99] Coward-Kelly G, Agbogbo FK, Holtzapple MT. Lime treatment of keratinous materials for the generation of highly digestible animal feed: 2. Animal hair. Bioresource Technology. 2006;**97**(11):1344-1352

[100] Chojnacka K, Gorecka H, Michalak I, Gorecki H. A review: Valorization of keratinous material. Waste and Biomass Valorization. 2011;**2**(3):317-321

[101] Tiwary E, Gupta R. Rapid conversion of chicken feather to feather meal using dimeric keratinase from *Bacillus licheniformis* ER-15. Journal of Bioprocessing & Biotechniques. 2012;**2**:123

[102] Brandelli A, Salab L, Kalil SJ. Microbial enzymes for bioconversion of poultry waste into added-value products. Food Research International. 2015;**73**:3-12

[103] Staron P, Banach M, Kowalski Z, Staron A. Hydrolysis of keratin materials derived from poultry industry. Proceedings of ECOpole. 2014;**8**(2):443-448

[104] Kornillowicz-Kowalska T, Bohacz J. Biodegradation of keratin waste: Theory and practical aspects. Waste Management. 2011;**31**(8):1689-1701

[105] Singh I, Kushwaha RKS. Keratinases and microbial degradation of keratin. Advances in Applied Science Research. 2015;**6**(2):74-82

[106] Riffel A, Lucas FS, Heeb P, Brandelli A. Characterization of a new keratinolytic bacterium that completely degrades native feather keratin. Archives of Microbiology. 2003;**179**(4):258-265

[107] Anbu P, Hilda A, Sur HW, Hur BK, Jayanthi S. Extracellular keratinase from Trichophyton sp. HA-2 isolated from feather dumping soil. International Biodeterioration & Biodegradation. 2008;**62**(3):287-292

[108] Kansoh AL, Hossiny EN, Hameed EK. Keratinase production from feathers wastes using some local Streptomyces isolates. Australian Journal of Basic and Applied Sciences. 2009;**3**(2):561-571

[109] Brandelli A, Daroit DJ, Riffel A. Biochemical features of microbial keratinases and their production and applications. Applied Microbiology and Biotechnology. 2010;**85**(6): 1735-1750

[110] Revathi K, Viruthagiri T. Optimization of process parameters for keratinase enzyme production using statistical experimental design. International Journal of Engineering and Innovative Technology. 2016;**6**(1):14-21

[111] Purchase D. Microbial keratinases: Characteristics, biotechnological applications and potential. In: Gupta VK, Sharma GD, Tuohy MG, Gaur R, editors. The Handbook of Microbial Bioresources. Wallingford, UK: CAB International Publishing; 2016. pp. 634-674

[112] Govarthanan M, Selvankumar T, Selvam K, Sudhakar C, Aroulmoji V, Kamala-Kannan S. Response surface methodology based optimization of keratinase production from alkali-treated feather waste and horn waste using Bacillus sp. MG-MASC-BT. Journal of Industrial and Engineering Chemistry. 2015;**27**:25-30

[113] Sharma A, Chandra S, Sharma M. Difference in keratinase activity of dermatophytes at different environmental conditions is an attribute of adaptation to parasitism. Mycoses. 2012;**55**(5):410-415

[114] Kublanov IV, Tsiroulnikov KB, Kaliberda EN, Rumsh LD, Haertle T, Bonch-Osmolovskaya EA. Keratinase of an anaerobic thermophilic bacterium Thermoanaerobacter sp. strain 1004-09 isolated from a hot spring in the Baikal rift zone. Microbiology. 2009;**78**(1):67-75

[115] Tork SE, Shahein YE, El-Hakim AE, Abdel-Aty AM, Aly MM. Purification and partial characterization of serine-metallokeratinase from a newly isolated Bacillus pumilus NRC21. International Journal of Biological Macromolecules. 2016;**86**:189-196

[116] Suzuki Y, Tsujimoto Y, Matsui H, Watanabe K. Decomposition of extremely hard-to-degrade animal proteins by thermophilic bacteria. Journal of Bioscience and Bioengineering. 2006;**102**(2):73-81

[117] Marchisio VF. Keratinophilic fungi: Their role in nature and degradation of keratinic substrates. In: Kushwaha RKS, Guarro J, editors. Biology of Dermatophytes and Other Keratinophilic Fungi. Bilbao: Revista Iberoamericana de Micologıa; 2000. pp. 86-92

[118] Blyskal B. Fungi utilizing keratinous substrates. International Biodeterioration & Biodegradation. 2009;**63**(6):631-653

[119] Soomro IH, Kazi YF, Zardari M, Shar AH. Isolation of keratinophilic fungi from soil in Khairpur city, Sindh, Pakistan. Bangladesh Journal of Microbiology. 2007;**24**(1):79-80

[120] Ajello L. The dermatophyte, *Microsporum gypseum*, as a saprophyte and parasite. Journal of Investigative Dermatology. 1953;**21**(3):157-171

[121] Lee MJ, Park JS, Chung H, Jun JB, Bang YJ. Distribution of soil keratinophilic fungi isolated in summer beaches of the east sea in Korea. Korean Journal of Medical Mycology. 2011;**16**(2):44-50

[122] Sharma M, Sharma M. Incidence of dermatophytes and other keratinophilic fungi in the schools and college playground soils of Jaipur, India. African Journal of Microbiology Research. 2010;**4**(24):2647-2654

[123] Rizwana H, Abdulaziz A, Hazzani A, Siddiqui I. Prevalence of dermatophytes and other keratinophilic fungi from soils of public parks and playgrounds of Riyadh, Saudi Arabia. The Journal of Animal & Plant Sciences. 2012;**22**(4):948-953

[124] Kumar R, Mishra R, Maurya S, Sahu B. Prevalence of keratinophilic fungi in piggery soils of Jharkhand, India. An International Quarterly Journal of Environmental Science. 2012;**1**:93-98

[125] Malek E, Moosazadeh M, Hanafi P, Nejat ZA, Amini A, Mohammadi R, Kohsar F, Niknejad F. Isolation of keratinophilic fungi and aerobic actinomycetes from park soils in Gorgan, north of Iran. Jundishapur Journal of Microbiology. 2013;**6**(10):e11250

[126] Tambekar DH, Mendhe SN, Gulhane SR. Incidence of dermatophytes and other keratinolytic fungi in the soil of Amravati (India). Trends in Applied Sciences Research. 2007;**2**(6):545-548

[127] Kumawat TK, Sharma V, Seth R, Sharma A. Diversity of keratin degrading fungal flora in industrial area of Jaipur and keratinolytic potential of *Trichophyton mentagrophytes* and *Microsporum canis*. International Journal of Biotechnology and Bioengineering Research. 2013;**4**(4):359-364

[128] Sangali S, Brandelli A. Feather keratin hydrolysis by a Vibrio sp. strain kr2. Journal of Applied Microbiology. 2000;**89**(5):735-743

[129] Lucas FS, Broennimann O, Febbraro I, Heeb P. High diversity among feather-degrading bacteria from a dry meadow soil. Microbial Ecology. 2003;**45**(3):282-290

[130] Ramnani P, Singh R, Gupta R. Keratinolytic potential of *Bacillus licheniformis* RG1 structural and biochemical mechanism of feather degradation. Canadian Journal of Microbiology. 2005;**51**(3):191-196

[131] Sivakumar T, Shankar T, Ramasubramanian V. Purification properties of *Bacillus thuringiensis* TS2 keratinase enzyme. American-Eurasian Journal of Agricultural & Environmental Sciences. 2012;**12**(12):1553-1557

[132] Riffel A, Brandelli A. Keratinolytic bacteria isolated from feather waste. Brazilian Journal of Microbiology. 2006;**37**(3):395-399

[133] Vigneshwaran C, Shanmugam S, Kumar TS. Screening and characterization of keratinase from *Bacillus licheniformis* isolated from Namakkal poultry farm. Research. 2010;**2**(4):89-96

[134] Vanbreuseghem R. Technique biologique pour l'isolement des dermatophytes du sol. Annales de la Societe Belge de Medecine Tropicale. 1952;**32**:173-178

[135] Otcenasek M. Ecology of dermatophytes. Mycopathologia. 1978;**65**(1-3):67-72

[136] Kumawat TK, Sharma A, Bhadauria S. Influence of liquid culture media, temperature and hydrogen ion concentration on the growth of mycelium and sporulation of *Arthroderma multifidum*. International Journal of Pharmaceutical Sciences Review and Research. 2016;**41**(2):136-141

[137] Zhao H, Huang L, Xiao CL, Liu J, Wei J, Gao X. Influence of culture media and environmental factors on mycelial growth and conidial production of *Diplocarpon mali*. Letters in Applied Microbiology. 2010;**50**(6):639-644

[138] Kumawat TK, Sharma A, Bhadauria S. Effect of culture media and environmental conditions on mycelium growth and sporulation of *Chrysosporium queenslandicum*. International Journal of ChemTech Research. 2016;**9**(11):271-277

[139] Sharma A, Sharma M, Chandra S. Influence of temperature and relative humidity on growth and sporulation of some common dermatophytes. Indian Journal of Fundamental and Applied Life Sciences. 2012;**2**(4):1-6

[140] Gopinath SCB, Anbu P, Lakshmipriya T, Tang TH, Chen Y, Hashim U, Ruslinda AR, Md Arshad MK. Biotechnological aspects and perspective of microbial keratinase production. BioMed Research International. 2015. Article ID: 140726. 10 p. DOI: 10.1155/2015/140726

[141] Williams CM, Richter CS, MacKenzie JM, Shih JCH. Isolation, identification, and characterization of a feather-degrading bacterium. Applied and Environmental Microbiology. 1990;**56**(6):1509-1515

[142] Cao ZJ, Zhang Q, Wei DK, Chen JW, Zhang XQ, Zhou MH. Characterization of a novel *Stenotrophomonas* isolate with high keratinase activity and purification of the enzyme. Journal of Industrial Microbiology and Biotechnology. 2009;**36**(2):181-188

[143] Umedum CU, Ugorji IR, Okoye EC, Egemonye OC, Obiora SO. Keratinase activity and biodegradation properties of hyphomycetous fungi from fowl feather. International Journal of Agriculture and Biosciences. 2013;**2**(6):306-309

[144] Adelere IA, Lateef A. Keratinases: Emerging trends in production and applications as novel multifunctional biocatalysts. Kuwait Journal of Science. 2016;**43**(3):118-127

[145] Mini KD, Paul MK, Mathew J. Screening of fungi isolated from poultry farm soil for keratinolytic activity. Advances in Applied Science Research. 2012;**3**(4):2073-2077

[146] Nigam PS. Microbial enzymes with special characteristics for biotechnological applications. Biomolecules. 2013;**3**(3):597-611